本书献给广东省建筑设计研究院（GDAD）65周年院庆
To the 65th Anniversary of GDAD

持守本源 筑梦千里
DESIGN AND BUILD FOR DREAMS

广东省建筑设计研究院（GDAD）65周年作品集
THE 65TH ANNIVERSARY COLLECTION OF GDAD

主编：陈雄 江刚 等
Editors-In-Chief: Chen Xiong Jiang Gang et al

中国建筑工业出版社
CHINA ARCHITECTURE & BUILDING PRESS

Foreword 序言

曾宪川

GDAD党委书记
高级建筑师
国家注册城市规划师

65年沧海桑田。回首望去，在中国的南粤大地上，有这么一群充满活力、敬业、专注、实干、求新的设计师，始终坚守在祖国城乡建设的前沿阵地，脚踏实地、孜孜不倦、披星戴月、栉风沐雨、兢兢业业、勤勤恳恳，秉承艰苦卓绝的创业和创新精神，初心不改，代代相传……

1952年用手摇计算机，计算出楼板、次梁、主梁、柱、墙的荷载以及配筋，再用丁字尺、三角板、鸭嘴笔、绘图板，手工绘制出构件、立面、剖面图，完成大批工业建筑设计，他们说，"保证完成任务"；

1957年挑着手摇钻探机、脚架、标尺、被袋、水桶、蚊帐、草席，天天爬山坡串田野，完成大批城镇勘探测量，他们说，"全心全意为人民服务"；

1965年深入粤北山区服务三线工程，边设计边施工，打穿山洞、运送余泥、浇灌混凝土，他们说，"坚决响应党中央号召"；

1975年走出国门援建第三世界国家，完成塞内加尔共和国5万观众席的体育场设计，他们说，"光荣啊，为国争光"；

1982年第一批进驻深圳经济特区，瓦砾遮头、风餐露宿，完成第一座图书馆、第一栋高层建筑、第一个路网规划等9项深圳第一的设计创举，他们说，"引领深圳速度"；

1990年完成当时国内第一高楼广东国际大厦的设计任务，首创采用筒中筒结构和无粘结部分预应力大跨度平板结构，节省投资、争取最大楼层净高，他们说，"要做敢为天下先的排头兵"；

2004年经过1600多个日夜鏖战，完成广州白云国际机场一号航站楼这项世纪工程，擦亮中国的南大门，他们说，"只有被追赶时才能跑得更快"；

2017年，这群人正带领着一支全专业高素质的技术服务团队，转型创新，向着一个又一个复杂高要求项目发起攻坚，他们说，"只有主动作为才能大有可为"。

这群人中有可敬的院士大师，有优秀的专业领军人物、生产经营能手，更有一大批默默奉献的一线设计师。他们殚精竭虑、废寝忘食、日夜兼程，他们用设计刻录时光，用坚持攀越高峰，用奋斗创造辉煌，他们在服务城乡建设的道路上始终专注地走了65年，见证了广东城乡的巨大变化，成就了今日之广东省建筑设计研究院，一个全球低碳城市和建筑发展倡议单位、全国科技先进集体、全国优秀勘察设计企业、当代中国建筑设计百家名院、全国企业文化建设示范单位、广东省高新技术企业、广东省守合同重信用企业、广东省抗震救灾先进集体、广东省重点项目建设先进集体。

然而这群人并没有停下前行的脚步，持续坚持以创业的激情、创新的思维、创造的实践和先行先试的勇气，不断开拓建筑工程设计、市政行业设计、工程勘察、工程咨询、城乡规划编制、建筑智能化系统工程设计、风景园林工程设计、建筑装饰设计、工程建设监理、招标代理、工程承包、施工图审查等工程业务，立足广东、面向国内外开展现代技术服务，足迹覆盖广东省内21个地级市、国内大多数省份和直辖市，以及西藏、新疆、广西等自治区、澳门特别行政区。

这群人坚守着"适用、经济、绿色、美观"的建筑方针，探索着自然与人的共生关系，引导着历史人文与工程项目的有机融合；坚持服务品质和质量，恪守行为准则和规范，为城乡建设提供一体化的技术解决方案；这群人骨子里有着现代工匠的坚持、包容务实的胸怀和对"守正鼎新、营造臻品"的坚定信仰。

宝剑锋从磨砺出，梅花香自苦寒来。值此GDAD成立65周年之际，从这群人的代表作品中精选部分汇编成册，记录65年来的实践征程，以为纪念，向他们致敬；同时传播他们的声音，发扬他们的传统，共同为现代城乡建设"添砖加瓦"。

Zeng Xianchuan

Party Secretary, GDAD
Senior Architect
Registered Urban Planner

Time flies and 2017 marks the 65th anniversary of GDAD. Looking back, we can see a dynamic and ambitious designer team in south China who has been working ceaselessly to promote the urban and rural development in China. Their hard-working, down-to-earth, enterprising and innovative spirit have been and are still being passed from one generation to another.

In 1952, with mechanical calculator, they calculated load and the reinforcement of the floors, secondary beams, main beams, columns and walls; with T-square, triangular plate, duckbill pen, and drawing board, they drew the components, elevations and sections manually, completin gdesigns for a large number of industrial buildings. They said, "we will accomplish the task".

In 1957, carrying hand drilling machines, tripods, rulers, sleeping bags, buckets, mosquito nets and straw mats, they climbed the hillside and walked across the farmland every day to accomplish a great number of urban surveys. They said, "we will serve the people with all our heart and soul".

In 1965, they participated in the Third Front projects in the northern mountainous area of Guangdong Province, designing at site while assisting with tunneling, debris handling and concrete-pouring. They said, "we resolutely respond to the call of the CPC".

In 1975, they went abroad to aid the third-world countries, completing the design of 50,000-seating stadium in the Republic of Senegal. They said, "we are the proud representatives of China".

In 1982, being the first to station in Shenzhen Special Economic Zone and facing hard living conditions, they completed the design of the first library, the first high-rise building, the first road network planning and etc. in Shenzhen. They said, "we must catch up the Shenzhen speed".

In 1990, they completed the design of the then highest building in China, Guangdong International Hotel, using the tube-in-tube structure and non-bonded partially-prestressed large-span slab structure for the first time, which cut the costs and maximizes the net floor height. They said, "we are never afraid of trying".

In 2004, after more than 1,600 days and nights of hard work, they completed the design of Terminal 1 of Guangzhou Baiyun International Airport, a trans-century project, widening China's south gate. They said, "you will only run faster when being run after".

Now in 2017, they are leading a multi-disciplinary technical team to meet the challenges of the large complicated projects one after another. They said, "Initiatives is the way to achievement".

They are respectable academicians of Chinese Academy of Sciences, outstanding leading figures in professional fields, operation specialists, and a large number of front-line designers who make silent contribution. Passing numerous sleepless nights, they have endeavored to fill the time with design, to climb higher with perseverance, to create prosperity with strenuous efforts. 65 years of dedicated services in urban and rural construction has made them witnesses of the great transformation in urban and rural areas in Guangdong. They are the ones that make GDAD what it is today: a signatory of the Initiative for Global Low-carbon City and Architectural Development (China) that receives numerous honorable titles of National Technologically Advanced Collective, National Excellent Exploration and Design Enterprise, Top 100 Architectural Design Institutes in Contemporary China, National Model for Corporate Culture Building, New and Hi-tech Enterprise of Guangdong Province, Trustworthy Enterprise of Guangdong Province, Advanced Collective of Guangdong Province for Earthquake Relief Work, Advanced Collective of Guangdong Province for Key Project Construction.

However, they have never stopped moving forward. With the same entrepreneurial passion, innovative thinking, creative practice and the courage of pioneers, they have been promoting their services of architectural engineering design, municipal utility design, engineering survey, engineering consulting, urban and rural planning, building intelligent system engineering landscape architecture, architectural decoration design, construction supervision, bidding agency, project contracting, and construction drawing review. Based in Guangdong and providing modern technical services at home and abroad, GDAD has set foot in 21 prefecture-level cities in Guangdong, most of the provinces and municipalities directly under the central government, autonomous regions including Tibet, Xinjiang, and Guangxi, as well as Macao Special Administrative Region.

Adhering to the principle of designing "suitable, cost-effective, green and aesthetical" buildings, they have been exploring the co-existence of nature and human beings and integration of culture with the engineering projects. Prioritizing the service quality and sticking to the code of conduct and norms, they have offered integrated technical solution for urban and rural development. They are artisans in modern times, firmly believing in perseverance, inclusiveness and the values of integrity, responsibility, innovation and development.

Good honing gives a sharp edge to a sword, and bitter coldness adds keen fragrance to plum blossom. On the occasion of the 65th anniversary of GDAD, we select some of the representative works to compile a book as a remembrance of the practice in the past 65 years and a tribute to our people. At the same time, this book also speaks for them and carries forward their efforts to promote the modern urban and rural construction.

Contents 目录

005	序言	Foreword
015	作品集萃	Selected Works
016	**文化、会展建筑**	**Cultural and Expo Buildings**
018	广东省博物馆新馆	New Guangdong Museum
022	广州市国家档案馆一期	National Archives of Guangzhou (Phase I)
024	南方广播影视创意基地（一期）	Creative Base (Phase I) of Southern Media Corporation
026	河源恐龙博物馆	Dinosaur Fossils Museum, Heyuan
028	粤剧艺术博物馆	Cantonese Opera Art Museum
032	泰山会展中心	Taishan Convention & Exhibition Center
036	广东（潭洲）国际会展中心首期工程	Guangdong (Tanzhou) International Convention and Exhibition Center (Phase I)
040	**科技教育建筑**	**S & T and Academic Buildings**
042	北京师范大学教学办公楼	Teaching & Office Building of Beijing Normal University
044	珠海翔翼保税区项目	Project in Zhuhai Xiangyi Bonded Area
046	广州气象雷达站	Guangzhou Weather Radar Station
050	ADG·机场设计研究院办公楼	ADG Office Building
054	华润大学	China Resources University
058	正佳海洋世界生物馆	Grandview Aquarium
060	横琴创意谷	The Inno Valley Hengqin
062	**商业、办公及酒店建筑**	**Commerce, Office and Hospitality Buildings**
064	索菲特酒店(圣丰广场)	Sofitel Hotel (Shengfeng Plaza)
068	广东全球通大厦	Global Access Building
070	江门电视中心	Jiangmen TV Center
072	广东广播中心	Guangdong Broadcasting Center
074	广州正佳商业广场东塔、西塔	East/West Tower of Grand View Plaza, Guangzhou
076	深圳万象城	The MixC, Shenzhen
080	广州香格里拉大酒店	Shangri-La Hotel, Guangzhou
082	惠州华贸中心商场	Huamao Center Mall, Huizhou
086	东莞市商业中心区F区（海德广场）	Dongguan Commercial Center Zone F (Haide Plaza)
090	从化新城市民之家	Conghua New City Citizen Center
092	广州气象卫星地面B站区业务楼	Business Building of Guangzhou Meteorological Satellite Ground Station Zone B
096	保利地产·珠海横琴发展大厦	Poly·Zhuhai Hengqin Development Mansion
100	美林湖度假酒店	Mayland Resort Hotel
102	海上世界酒店	Sea World Hotel

106	白云绿地金融中心	Baiyun Greenland Financial Center
110	招商局广场	China Merchants Group Plaza
114	中广核大厦	CGN Building
118	海上世界广场-船后广场	Seaworld Plaza – Ship-back Square
122	海上世界广场-船前船尾广场	Seaworld Plaza – Ship-front Square and Ship-end Square
124	昆明西山万达广场-双塔	Xishan Wanda Plaza – Super High-rise Office (Twin Towers)
128	珠江新城F2-4地块项目	Plot F2-4, Zhujiang New Town, Guangzhou
132	保利商务中心	Poly Business Center
136	**交通建筑**	**Transportation Buildings**
138	广州新白云国际机场一号航站楼	Terminal 1, Guangzhou New Baiyun International Airport
142	广州新白云国际机场二号航站楼	Terminal 2, Guangzhou New Baiyun International Airport
146	揭阳潮汕机场航站楼及配套工程	Jieyang Chaoshan Airport Terminal and Supporting Works
150	深圳机场卫星厅	Shenzhen Airport Satellite Concourse
152	深圳机场新航站区地面交通中心（GTC）	Ground Transportation Center (GTC) of New Terminal Area, Shenzhen Airport
154	深圳蛇口邮轮中心	Shekou Cruise Center, Shenzhen
158	武汉火车站	Wuhan Railway Station
162	新建云桂铁路引入昆明枢纽工程 新建昆明南站	Kunming South Railway Station (New)
166	广州市轨道交通四号线车陂南-黄阁段工程	Metro Stations of Guangzhou Metro Line 4
168	广州市轨道交通五号线首期工程-坦尾站	Tanwei Station of Guangzhou Metro Line 5
170	广州市轨道交通五号线工程-动物园站	Zoo Station of Guangzhou Metro Line 5
172	**体育建筑**	**Sports Buildings**
174	广州亚运馆	Guangzhou Asian Games Gymnasium
178	2008年奥运会北京老山自行车馆	Laoshan Velodrome for 2008 Beijing Olympic Games
180	广州自行车馆	Guangzhou Velodrome
184	佛山市岭南明珠体育馆	Lingnan Pearl Gymnasium, Foshan
188	惠州市金山湖游泳跳水馆	Jinshan Lake Swimming and Diving Natatorium, Huizhou
192	广州市花都区东风体育馆	Dongfeng Gymnasium, Huadu District, Guangzhou
196	肇庆新区体育中心	Zhaoqing New Area Sports Center
198	惠州博罗县体育中心体育场	Sports Center Stadium, Boluo County, Huizhou
200	**医疗建筑**	**Medical Buildings**
202	中山大学肿瘤防治中心	Sun Yat-Sen University Cancer Center
204	株洲市中心医院	Zhuzhou Central Hospital
206	梅州市人民医院	The People's Hospital of Meizhou City
208	惠州市妇幼保健院	Maternal and Child Health Hospital of Huizhou City
210	海螺医院	Conch Hospital

212	**地下空间**	**Underground Spaces**
214	广州市珠江新城核心区市政交通项目	Municipal Transportation Project in Core Area of Zhujiang New Town, Guangzhou
218	金融城地下空间	Underground Space of Guangzhou International Financial City
220	万博地下空间	Underground Space of Wanbo Business District
222	**居住建筑**	**Residential Buildings**
224	广州猎德村旧村改造项目	Reconstruction of Liede Village, Guangzhou
226	广州招商金山谷花园(1- 4期)	The Hills (Phase I- IV), Guangzhou
228	深圳招商美伦公寓	Maillen Apartment, Shenzhen
232	广州科学城科技人员公寓	S & T Professionals' Apartment, Guangzhou Science City
236	佛山依云水岸A1、B1、B2项目	Project A1, B1 and B2 of Evian Town, Foshan
238	香港新福港地产·广州萝岗鼎峰	SFK · DF Project, Luogang, Guangzhou
240	招商伍兹公寓	China Merchants Woods Park
242	鲸山花园九期	Jingshan Garden Phase IX
246	万科科学城项目—— 一期工程	Vanke Project in Guangzhou Science City (Phase I)
250	金隅大成·海口美灵湖住宅小区	Jinyu Dacheng- Meiling Lake Residential District, Haikou
252	金山谷花园3B期、八期	The Hills Phase III-B and Phase VIII
254	华润小径湾花园(一期住宅及商业)	China Resources Xiaojing Bay Garden (Phase I Residence and Retails)
258	南沙港航华庭	Ganghang Huating Complex, Nansha
262	**规划设计**	**Planning and Design Urban**
264	中新广州知识城主城区城市设计深化及控制性详细规划	Urban Design Detailing and Regulatory Detailed Plan of Main Urban Area of Sino-Singapore Guangzhou Knowledge City
266	萝岗中心区·广州科学城城市设计及景观规划	Urban Design and Landscape Planning of Luogang Central Area · Guangzhou Science City
268	南沙新区蕉门河中心区中区和南区环境整治与景观提升设计总咨询	Consultancy for Environmental and Landscape Improvement of Jiaomen River Central Area (South/Middle Zone), Nansha
270	广州番禺万博商务区地下空间控制性详细规划	Regulatory Detailed Plan of Underground Space of Wanbo CBD, Guangzhou
272	**市政桥道**	**Bridges and Roads**
274	广州铁路新客站地区市政及附属工程	Municipal and Ancillary Works in New Guangzhou Railway Passenger Station Area
276	番禺南大干线	Panyu South Main Line
278	华南路三期工程	Huanan Expressway (Phase III)

280	**市政给排水**	**Municipal Water Supply & Drainage**
282	广州市西江引水工程	Xijiang River Water Diversion Project, Guangzhou
284	广州市花都区新华污水处理厂（一、二期）提标改造工程	Xinhua Sewage Treatment Plant (Phase I and II) Upgrading and Renovation Project in Huadu District, Guangzhou
286	广州市江高-石井污水处理系统工程	Guangzhou Jianggao – Shijing Sewage Treatment System
288	**水处理与环保技术**	**Water Treatment and Environmental Protection**
290	花都水厂工程	Huadu Water Plant
292	江门市区应急备用水源管道及供水设施工程	Jiangmen Urban Emergency Backup Water Supply Pipelines and Facilities
294	郁南县整县生活污水处理捆绑PPP项目	PPP Bundled Project for Sewage Treatment Plants of Yunan County
296	**环境保护**	**Environmental Protection Projects**
298	台山市台城下豆坑生活垃圾卫生填埋场一期工程	Xia Dou Keng Domestic Waste Landfill (Phase I), Taicheng, Taishan
300	惠州市惠阳区榄子垅环境园生活垃圾综合处理场BOT项目	Lan Zi Long Environmental Park Domestic Waste Treatment Plant (BOT), Huizhou
302	佛山高明苗村白石坳垃圾卫生填埋场渗滤液处理厂二期工程	Leachate Treatment Plant (Phase II), Bai Shi Ao Waste Sanitary Landfill, Foshan
304	**景观设计**	**Landscape Design**
306	广州铁路新客站地区公共绿化和广场景观工程	Greening and Square of New Guangzhou Railway Station, Guangzhou
308	珠海玲玎海岸园林景观及绿化设计	Lingding Coastal Landscape and Greening Design, Zhuhai
310	广州市金沙洲居住新城P线景观、K、M、N、U路绿化工程	Jin Sha Zhou Line P Landscaping and Roads K/M/N/U Greening Project, Guangzhou
312	**室内装修**	**Interior Design**
314	广州报业文化中心	Guangzhou Daily Group Culture Center
316	广州民俗博物馆	Guangdong Folk Arts Museum
318	中国南方航空大厦室内装饰装修工程设计	Interior Fit-out for China Southern Airlines Building, Guangzhou
320	南方医院惠侨楼室内装修工程设计	Interior Fit-out for Huiqiao Building, Nanfang Hospital, Guangzhou
323	**经典作品**	**Classic Works**
333	**后记**	**Epilogue**

Selected Works 作品集萃

文化、会展建筑

文化活动是人类重要的社会行为之一，是人类文明进步的必要条件。文化建筑给人们提供了社会文化活动的建筑、环境空间，是公共建筑的重要组成部分。文化建筑涵义宽泛，展览、演示、收藏、研究人类社会文化、生活和生产成就，是这一类型建筑的主要功能，建筑类型包括了博物馆、美术馆、艺术中心、展览中心等。在建筑艺术领域中，文化建筑缘于其功能和空间特点，是建筑师创作的重点对象，也是彰显文化特点、文化成就和艺术形象的标志。

在超过半个世纪的建筑设计实践中，GDAD设计了大量的文化建筑，绝大部分都成了地方建筑艺术和城市形象的代表，为社会贡献了一批珍贵的建筑财富。在把握文化建筑的本体历史形象、地方环境特点、社会民俗风貌、空间构成意念等方面，GDAD的设计作品都作出了行之有效的探索，总结了大量的设计经验，反映在本图册里展示的部分近期代表作之中。

文化建筑设计，首先需要把握的，是建筑的时代特点。建筑存在于历史，更应呈现于当下。反映时代的意念流向、技术进步，是建筑历史存在性的必然。广东省博物馆新馆用悬挂式钢结构营造收藏艺术珍品的宝盒，梅州客家艺术中心以现代设计手法和结构体系，体现客家传统围龙屋的意念，河源恐龙博物馆将远古菊石的形象抽象为现代建筑造型，都对建筑历史性与时代性的结合作出了回应。

地方的就是世界的，虽为老生常谈，但在于文化建筑，却有其实际意义。文化建筑的地方历史文化特点，确定着建筑的本体内涵。广东省博物馆新馆用漆器的隐喻来彰显岭南文化，泰山会展中心将祭天金鼎的片段浓缩为建筑母题来表达对齐鲁文化的尊敬，广州市国家档案馆一期将岭南花窗构成于建筑立面中以体现岭南传统的表征。这些设计手段，将传统文化的亲和力赋予当代建筑，增强了建筑的表现力。

重视建筑的使用功能，使用现代技术手段，将建筑空间的表现力与功能有机结合，这是文化建筑的重要设计技巧。南方广播影视创意基地需要将多种不同使用功能和不同体量要求的演艺传媒空间组合在建筑中，设计上巧妙地结合地形，对各类功能用房结合建筑形象进行有机组合，建筑形象与使用空间达到和谐统一的目的。粤剧艺术博物馆的建设需求是传统岭南园林与博物馆展厅的结合，设计中用分轴布线、空间叠合的方式，将园林的楼阁建筑与表演、展览等大空间功能用房叠合、穿插在一起，建筑园林主体既体现了传统韵味，又能满足博物展览的要求。梅州客家艺术中心的多类演艺和教学用房，以及河源恐龙博物馆的古生物挖掘遗址和馆藏展厅等，也是通过建筑师的不懈努力，使多种建筑功能与空间做到有机结合，建筑形象与使用功能融合自如。

建筑承载了多种多样的需求，建筑设计的方法也应不拘一格。建筑创作最基本的理念，难以离开历史文化和地方特色，难以脱离根本使用需求和相应的技术支撑，建筑师的工作，其实最主要的是将各类相关理念、功能、技术和设施加以组织和协调。

Cultural and Expo Buildings

Cultural activities represent an essential part of the human social behaviors and a prerequisite for the progress of a civilization. Cultural buildings provide people with built spaces for social and cultural activities, making up an important part of public buildings. Cultural buildings have a broad-sense definition. They mainly serve to exhibit, display, collect and study the culture, life and production achievements of human society. In terms of building typology, they include museums, art galleries, art centers and exhibition centers. In the field of architectural art, due to their distinctively featured functions and spaces, cultural buildings have been an focus for architects' creation and the symbols highlighting the cultural characteristics, cultural achievements and artistic image.

Following more than half a century of architectural design practices, GDAD has designed a great number of cultural buildings, the vast majority of which have become the landmarks representing the local architectural art and city image, contributing considerable architectural assets to the community. GDAD has effectively and extensively explored and summarized how to grasp and deliberate the historical image of the cultural buildings, the characteristics of local environment, the folk styles, and the spatial composition, as evidenced in its design works, the most recent representatives of which are compiled in the book.

First and foremost, the design of cultural buildings must reflect the features of the times. Buildings are part of the history, but more importantly, a reflection of the present. Reflecting the trend of the times and technological progress are inevitable for the historical nature of buildings. New Guangdong Museum uses a suspension steel structure to create the image of a box of treasures. Hakka Arts Centre, Meizhou adopts modern design techniques and structural system to embody the idea of the traditional Hakka walled village. Dinosaur Fossils Museum, Heyuan imitates the image of Ammonitida in ancient times to build a modern building. These projects all respond to the integration of the historical nature and the feature of times of architecture.

What is local is universal. Although often being talked of, it actually matters in the design of cultural buildings. The local historical and cultural characteristics of cultural buildings determine the connotation of the buildings. New Guangdong Museum uses lacquerware as a metaphor to highlight the Lingnan culture. Taishan Convention & Exhibition Center mirrors the *Ding* used for heaven worshipping ceremonies in ancient times to pay tribute to the ancient Qi Lu Culture. National Archives of Guangzhou (Phase I) incorporates the Lingnan latticed window into the building facade to reflect the Lingnan tradition. These design approaches attach the affinity with traditional culture to contemporary buildings, enhancing the expressiveness of the buildings.

Using modern technical means to integrate the spatial expressiveness and the functions valued of the buildings is an important technique in designing cultural buildings. The Creative Base of Southern Media Corporation needs to accommodate a variety of functions and performance and media spaces of different requirements in the building. With the terrain taken into consideration, various types of functional rooms are configured corresponding to the architectural appearance, realizing harmony and unity between the architectural appearance and the functional spaces. The Cantonese Opera Art Museum needs to combine the traditional Lingnan gardens with the exhibition halls. An axis-based layout with overlapped spaces are employed to stack and interweave the pavilions in the gardens and the large functional rooms for performance and exhibition. In this way, the gardens and buildings not only embody the traditional charm, but also meet the requirements for exhibition. The various types of performance and teaching space in Hakka Arts Centre, Meizhou, as well as the ancient creatures' excavation sites and collection exhibition hall in Dinosaur Fossils Museum, Heyuan, are also the results of the tireless efforts of architects to integrate various architectural functions with space, and the building appearance with functionality.

Buildings serve different needs. Therefore, architectural design techniques vary. The fundamental concept of architectural design is inseparable from the history, culture and local characteristics, and must be based on the fundamental needs and relevant technical support. Therefore, the most important task of the architects is to properly organize and coordinate various related concepts, functions, technologies and facilities.

广东省博物馆新馆
New Guangdong Museum

广东省博物馆新馆是广东省标志性文化工程之一，是广东建设文化大省的重点项目。

博物馆新馆的造型仿佛一件雕通的宝盒，这一设计意念源于广东传统的工艺品——象牙球，博物馆的空间组织就像象牙球镂空的工艺，内部功能层层相扣，展厅、回廊、中庭与整体结构紧密结合，由内向外逐层展开，利用虚实变换的隔断吸引观众层层而进，功能流线自然而生，使形式和功能形成统一的有机整体。

博物馆新馆的展陈部分采用钢桁架悬吊式设计，由形体中间67.5米×67.5米轴线交点处巨型钢筋混凝土筒体，外悬出23米跨的大型空间钢桁架，悬吊二至三层楼面，形成钢筋混凝土剪力墙——钢桁架悬吊结构体系。展厅内形成一个大跨度无柱式空间，高度由5米至9米或20米不等，参观者从主入口处直接进入到中庭，上部外悬挂展厅便映入眼帘，而外部感觉整个宝盒如同飘浮在起伏的绿化广场上。

The New Guangdong Museum is one of the iconic cultural projects of Guangdong province and one of the key projects as a part of Guangdong's efforts to further enhance the culture power.

The project takes the shape of a hollow-out jewelry box, which is inspired by Cantonese traditional handcrafts: carved ivory ball. The space of the Museum is organized by referencing the craftsmanship of ivory ball. The internal functions are interconnected with exhibition halls, cloisters and atrium closely combined into the overall structure. The space unfolds inside-out layer by layer with alternating void and solid partition to arouse the visitor's appetite for exploration. This way, the functional circulation formed naturally, organically unifying the building form and function.

The exhibition complex of New Guangdong Museum features steel truss suspension structure. A large spatial steel truss of 23-m-span is overhung from the giant reinforced concrete core located at the intersection of 67.5m X 67.5m axes of the building form and suspended to the second and third floor to form reinforced concrete shear wall, i.e. steel truss suspension structure system. The structure allows a large span column-free space with floor height varying from 5 meters to 9 meters or 20 meters. As visitors access to the atrium from the main entrance, the overhead hanging exhibition halls appear. Externally, the whole building looks like a jewelry box suspended on the green square.

2

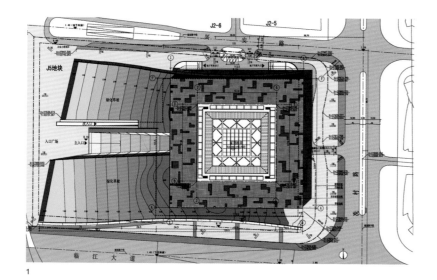

1

项目地点：广州市天河区珠江东路
设计时间：2003.7
建设时间：2005.11
建筑面积：66000m²
建筑层数：地上5层，地下1层
建筑高度：45.4m
合作单位：许李严建筑师有限公司
曾获奖项：2011年度全国优秀工程勘察设计行业奖 建筑工程一等奖
中国文化建筑范例工程
2011年度广东省优秀工程勘察设计奖一等奖

Location: Zhu Jiang Dong Lu, Tianhe District, Guangzhou
Design: 2003.7
Construction: 2005.11
GFA: 66,000m²
Floors: 5 aboveground, 1 underground
Height: 45.4m
Partner: Rocco Design Architects Limited
Awards: The First Prize of Architectural Engineering under National Excellent Engineering Exploration and Design Award, 2011
Model Project for Cultural Buildings in China
The First Prize of Excellent Engineering Design Award of Guangdong Province, 2011

1 总平面图
  Site plan

2 绿色绸缎盛托着雕通宝盒
  A hollow-out jewelry box suspended on green silk fabric

3 夜景效果
  Night view

4 别具心思的外墙由单双色铝单板和穿孔铝板，与玻璃和饰面屏风组成
  The creative exterior walls feature alternated single-color and double-color aluminum veneer, perforated aluminum sheet, glass and decorative screen

5 立面局部肌理
  Partial texture of facade

6 立面图
  Elevation

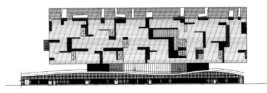

7

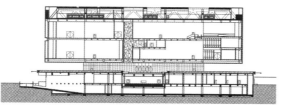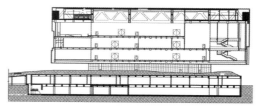

8

7 内部中庭高度从观众入口大厅直达顶层，自然光线由顶部玻璃天窗铺撒而下
Floor to ceiling internal atrium at the entrance hall allows the day light to flood indoors via the skylights at the top

8 剖面图
Section

9 走廊围绕中庭空间展开，将人的视线再次引入中庭
Cloisters extend around atrium and guide visitors' eyesight again back to the atrium

10 夹层空间的设计，给予不同的人群产生对话的可能
Mezzanine space makes it possible to have dialogue between different people

11/12 通高连廊带给观赏者庄严的气氛
Full-height cloisters convey an atmosphere of majesty

13 丝绸般内部设计时实时虚，纯粹大气
Silk-like internal design with contrast between void and solidness creates an atmosphere of purity and grandeur

广州市国家档案馆一期
National Archives of Guangzhou (Phase I)

广州市国家档案馆是广州市一项重要的文化设施。项目建设力争"四个一流"，即：规模一流、质量一流、功能一流和信息化一流，立足于服务广东，示范全国。2012年12月28日正式对市民开放。广州市国家档案馆用地面积为50013.2平方米，一期总建筑面积：28881平方米。按甲级档案馆设计，建筑设计考虑文化性、开放性和公众性。总平面布置紧凑合理，总体T字形布局与三角形用地巧妙结合，一期长方形造型，位于地块中部。建筑空间的变化使一期在面积不大的情况下，有一定的体量和规模，主导地位突出。一期建筑室内外空间错落变化。按岭南园林档案馆组织室内外空间，符合现代档案馆"开放""服务"的公建特征，建筑设有中庭、连廊、不同标高屋顶花园、平台、外廊、构架等，建筑景观、朝向、采光、通风良好，也提升建筑办公空间品质及使用舒适度，节能环保。

作为体量简洁、比例优美、现代感强的文化殿堂，一期建筑通透大气，立面肌理结合传统窗花和广式雕刻元素，进行变形与组合，形成简约的建筑肌理，彰显现代和传统，形式和功能的巧妙结合。建筑外表采用规整排列的白色框架，并有适当微差变化处理，形成了档案架般的造型，暗喻了建筑物的功能属性。平面布局流畅，使用便利。设置有对外服务用房、库房、业务与技术用房、办公及附属用房。功能分区明确，避免互相交叉干扰，方便对外服务、使用管理及库区安全保卫。一期公共和内部区域划分明确。展厅以中庭为核心设计，空间布局符合展览要求。空间业务技术和办公楼，采用院落式布局，各房间均可南北对流。

外墙采用加气混凝土墙体，幕墙玻璃采用Low-E中空玻璃6+12A+6，库房外墙采用加气混凝土双墙和中间40毫米挤塑聚苯板，更好地满足库房恒温恒湿要求。

As one of the important cultural facilities in Guangzhou, National Archives of Guangzhou is developed to reach "four first classes", namely: first class in size, in quality, in function and in information technology, serving Guangdong Province and setting an example nationwide. Opening to the public on December 28, 2012, the National Archives of Guangzhou occupies a land area of 50,013.2m^2 with GFA of 28,881m^2 for Phase I. It is designed as per the standard for Grade A archives with consideration to culture, openness and publicness. With compact and reasonable master layout, the T-shape layout well fits the triangle site with rectangular Phase I in the middle of the site. The variety of building space allows sufficient volume and size of Phase I despite of limited area to highlight its leading role. The staggered outdoor and indoor spaces of phase I are organized by referencing the style of Lingnan gardens, demonstrating the openness and public service of modern archives as a public building. Atriums, corridors, roof gardens of different elevations, terraces, verandahs and frameworks ensure favorable landscape, orientation, daylighting and ventilation and help to improve the quality of office space and user's comfort in an energy-efficient and environmental way.

As a concise, well proportioned, modern cultural building, the transparent and grand Phase I building incorporates the elements of traditional paper-cut window decoration and Cantonese-style carving in façade, which are transformed and combined to further develop concise architectural texture, integrating modern and traditional features, form and function. The building is clad with regularly arranged white frame with slight variation, as the image of archive shelves, suggesting the building function. The smooth planar layout ensures convenient use. Public service room, warehouses, academic and technical rooms, office and auxiliary rooms are designed with clear functional zoning to avoid interference and facilitate public service, operation and warehouse security. Public and internal zones are clearly defined in Phase I. Exhibition Hall is designed around atrium with the spatial layout suitable for exhibition. Technical and office building is designed in courtyard style layout to ensure natural ventilation of each room.

Aerated concrete wall is adopted for exterior wall and low-E insulated glass 6+12A+6 is adopted for glass façade. Aerated concrete wall and middle 40mm extruded polystyrene panel are adopted for warehouse exterior wall to better meet the constant temperature and humidity requirements of warehouse.

项目地点：广州大学城中心区西侧
设计时间：2009.7-2010.1
建设时间：2009.12-2012.8
建筑面积：28881m^2
建筑层数：地上7层，地下1层
建筑高度：32m
曾获奖项：广东省第二届岭南特色建筑设计奖银奖
　　　　　蓝星杯·第七届中国威海国际建筑设计大奖赛优秀奖
　　　　　第八届全国优秀建筑结构设计奖三等奖
　　　　　2013年度广东省优秀工程勘察设计奖三等奖

Location: In the west of Central Area of Guangzhou Higher Education Mega Center
Design: 2009.7-2010.1
Construction: 2009.12-2012.8
GFA: 28,881m^2
Floors: 7 aboveground, 1 underground
Height: 32m
Awards: Silver Prize of the Second Lingnan Feature Architectural Design Award of Guangdong Province
Excellence Design Award of Blue Star Cup - the Seventh China Weihai International Architectural Design Competition
The Third Prize of the Eighth National Excellent Building Structure Design Award
The Third Prize of Engineering Design under Excellent Engineering Exploration and Design Award of Guangdong Province, 2013

1 总平面图
　Site plan
2 主体形象及景观
　Main building and landscape
3 西立面视角
　View from west facade
4 屋顶平台及立面构建细部
　Roof terrace and detail of façade construction
5 南侧景观主轴，纯净简洁有韵律感，突显建筑主体的庄重、精致、典雅的文化气息
　The pure, concise and rhythmic main landscape axis in the south sets off the main building that presents a culture of dignity, delicacy and elegance
6 屋顶平台
　Roof platform
7 北立面局部视角
　Partial view from north facade

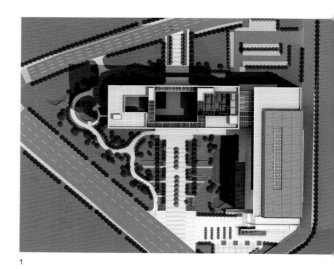

1

2 3 4

5

6 7

Cultural And Expo Buildings 文化、会展建筑

南方广播影视创意基地（一期）
Creative Base (Phase I) of Southern Media Corporation

随着广东建设文化强省的号角吹响，南方广播影视创意基地成为广东省建设文化强省的十大精品项目之一。第九届全运会及第十六届亚运会的成功举办，也推动了以奥体中心为核心的城市东部新区的发展。在环境优美的奥体公园北侧，一个集办公、影视、动漫节目制播、外景拍摄、旅游购物、培训展览、商务酒店等功能于一体的"中国好莱坞"将拔地而起。它将成为广州的新地标，彰显新世纪广东广播电视行业的新面貌。

本项目的方案设计必须满足原有的规划设计的基本框架。设计方案通过整合不同单体的体量、造型，统一建筑设计手法加强建筑群的整体感。整体交通规划利用地形自然形成两个相差6米的台地区分组织场地内外各类流线。内部功能设计必须满足集团现在及未来影视节目制作的要求，各类演播厅竖向组合，集中布置，内部工作流线便捷明晰。同时，空间设计也要考虑到参观或参与节目录制的公众，在保证功能合理布置及面积节约的前提下，充分满足公共空间的舒适性。酒店及会议中心设计根据其用地及功能特征，整合一体化，引入可方便组合联系的大堂及交通流线，两者可分可合资源共享。

建筑造型采用现代简洁的表达方式，引入非线性设计，产生丰富变幻的视觉效果。位于前广场的剧院造型更加具有创意及动感，动感的建筑造型带来了同样流动的室内空间。多媒体展厅结合参观流线形成独特的造型效果，别致的流线设计使人们可以由地面通过坡道自然地走到屋顶花园，带来新颖的参观体验。建筑中引入屋顶花园，采用自然地貌的设计手法，花园自然融入形体变化中，成为建筑的一部分，随处可达，提升了建筑环境品质。办公楼立面使用可呼吸式双层幕墙，根据需要调节入射光线及引入新风，兼顾了办公区的景观效果和空间舒适度要求。同时在立面设计中运用彩釉玻璃的不同透明度的设计组合，形成丰富的肌理变化，降低遮阳系数，节能环保。

In response to Guangdong's efforts in developing cultural powerhouse, the Creative Base of Southern Media Corporation is listed into the top ten quality projects for cultural development. The successes of the Ninth National Games and the Sixteenth Asian Games also drives the development in eastern new urban area around the Olympic Sports Center. Just to the north of the beautiful Olympic Sports Park, the "Chinawood" integrating the office, production and broadcasting of film, TV program and animations, location shooting, tourism, shopping, training, exhibition, business hotel and other functions springs from the ground, as a new landmark for Guangzhou and a new face of Guangdong TV and broadcasting industry in the new century.

The design must fit in the basic framework in the original planning. With integration of volumes and shapes of individual buildings, the unity of the building cluster is enhanced via consistent architectural design approach. As for the overall traffic plan, two terraces with a height difference of six meters are naturally shaped by following the terrain to organize internal and external circulations. The interior function must meet the present and future demands of TV and film production; various studios are vertically centralized to ensure a clear-cut and convenient internal working circulation. Meanwhile, the spatial design should manage to provide comfortable public space to the visitors and attendances for the program recording by ensuring rational functional layout and floor efficiency. The hotel and the convention center are integrated in view of the site and function features with lobby and traffic circulations for convenient connection, flexible separation and combination and resource sharing.

With modern and concise building shape, non-linear design is introduced for varied visual effects. The theatre at the front square assumes a more creative and dynamic look with flowing interior spaces. The multimedia exhibition hall presents unique appearance with the help of visiting circulation, which allows visitors to walk up to the roof garden from the ground via ramp, as a fresh visiting experience. The roof garden is introduced to and naturally integrated with the changing building form by following the natural landform, which is accessible everywhere as a part of the building for better building environment. The facades of office building employ breathable double-skin curtain walls for the daylight regulated and fresh air introduced as needed to realize both desired views and comfort level in the office area. At the same time, in the façade design, fritted glass with varied transparency are combined for diversified building fabrics, lowered shading coefficient and energy efficiency effect.

项目地点：广州天河
设计时间：2012
建设时间：2014年至今
建筑面积：544000m²
建筑层数：12层
建筑高度：50m
曾获奖项：2012年国内竞赛中标
2014年广东省注册建筑师协会第七次优秀建筑佳作奖

Location: Tianhe, Guangzhou
Design: 2012
Construction: 2014 to date
GFA: 544,000m²
Floors: 12
Height: 50m
Awards: Winning proposal of domestic competition in 2012
The Seventh Excellent Architecture Creation Award by Guangdong Chapter of Association of Chinese Registered Architects, 2014

1 总平面图
  Site plan

2 创意基地整体体量顺应地形，形成不同高差的两个台地，自然区分了内外广场
  The Base follows the terrain to form two terraces with different heights, naturally separating the internal and external squares

3 建筑立面根据朝向采取了不同的处理手法，达到生态节能的目标
  Different facade strategies are employed in view of orientation to respond to the ecological and energy goals.

4 剖面图
  Section

Cultural And Expo Buildings 文化、会展建筑 025

河源恐龙博物馆
Dinosaur Fossils Museum, Heyuan

河源恐龙博物馆是我国第一个以恐龙为主题的专业博物馆，建设地点位于河源市源城区龟峰公园内，环境优美，拟建基地位于宋代古塔——龟峰塔、金花庙及龟峰山的西南侧，建设范围内基本为平地，高程为40.0米。龟峰塔建于山坡顶，山坡顶高程60.0米。

项目采用螺旋上升的体量，把地形、景观、功能、空间有机结合。项目采用源于自然、依于景观的设计手法进行设计，博物馆的石材墙体由地面升起，舒展、盘回于山麓，与自然浑然一体，宛如天生。项目以"菊石传说中的龟峰山围墙"为设计构思，采用围合而又舒展打开的平面布局，如同苏醒的种子、新生的胚胎，承载着古老的基因，孕育着新生的文明。项目把远古的"菊石"形态作为建筑雏形，隐喻新生的胚胎，充分体现项目的文化性。

通过轴线设计，使建筑、景观、场地、保留建筑有机串联，并与整个龟峰公园形成整体。设置主入口轴线，形成具有清晰引导性、丰富变化性的空间序列，成为博物馆的参观主流线；设置博物馆与古塔的呼应轴线，使人们可以在大堂内部欣赏到古塔的全貌，并使龟峰塔成为博物馆永久展示的珍品；设置主入口广场与古塔的呼应轴线，形成人与建筑、人与自然的对话空间。合理规划交通系统，充分考虑参观流线、车行流线、后勤流线、布展流线及上山游览流线。合理布置各功能，充分考虑展厅、藏品库、办公、机房及展示大堂等功能的关系，做到既互相独立又联系便捷。

项目地点：广东省河源市
设计时间：2006.6-2006.11
建设时间：2006.11-2010.11
建筑面积：8196m²
建筑层数：地上3层，地下1层
建筑高度：15.9m
曾获奖项：2013年中国优秀文化建筑展览范例工程银奖
　　　　　2013年度全国优秀工程勘察设计行业奖三等奖
　　　　　2013年度广东省优秀工程勘察设计奖二等奖

Location: Heyuan City, Guangdong Province
Design: 2006.6-2006.11
Construction: 2006.11-2010.11
GFA: 8,196m²
Floors: 3 aboveground, 1 underground
Height: 15.9m
Awards: Silver Award of Model Projects for China's Excellent Cultural Building Exhibition, 2013
The Third Prize of National Excellent Engineering Exploration and Design Award, 2013
The Second Prize of Excellent Engineering Design Award of Guangdong Province, 2013

Located at the scenic Guifeng Park in Yuancheng District of Heyuan City, the Dinosaur Fossils Museum is the first dinosaur-themed specialized museum in China. On the southwest of Guifeng Pagoda (Song Dynasty), Golden Flower Temple and Guifeng Mountain, the site is flat with the altitude of 40m. The Guifeng Pagoda was built on the top of the slope (altitude of 60m).

The spiral-up volume is adopted to organically combine the terrains, landscapes, functions and spaces. With the design approach of having the Project grow out of nature and embedded in landscapes, stone walls rise straightly from the ground and wind along the mountain, looking like an innate part of the nature. Conceived from "Walled Guifeng Mountain in the Legend of Ammonite", the enclosed yet open planar layout, just like awakened seed and nascent embryo, carries the ancient genes and give birth to new civilization. The ancient "ammonite" as the architectural prototypes, metaphorizes new embryos and fully communicates its culture connotation.

The axes organically link up the building, landscape, site and preserved buildings to integrate with the Guifeng Park. A main entrance axis is provided to form a spatial sequence with clear orientation and varied changes, serving as the main touring route of the Museum. An axis between the Museum and the ancient Guifeng Pagoda allows visitors to have a full view of Guifeng Pagoda inside the lobby of the Museum and makes Guifeng Pagoda a permanent exhibit of the Museum. An axis between the main entrance square and ancient Guifeng Pagoda enables the dialogue between people and building, people and nature. The reasonably designed traffic system gives full consideration to circulations of museum visiting, vehicle, BOH, exhibition move-in and mountain climbing. The functions are reasonably distributed with due consideration to the relation among exhibition halls, collection storage offices, mechanical rooms and the exhibition lobby for both mutual independence and convenient connection.

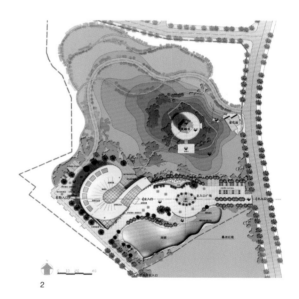

2

1 手绘图
　Sketches

2 总平面图
　Site plan

3 博物馆的石材墙体由地面升起，舒展、盘回于山麓，与自然浑然一体，宛如天生
　The stone walls of the Museum rise straightly from the ground and wind along mountain, as an innate part of the nature

4 河源恐龙博物馆采用螺旋上升的体量，与山体有机结合
　The Museum has a spiral-up volume organically integrated with the mountain

5 透过静谧的水面，倒影与建筑融为一体
　The reflection on the tranquil water surface and buildings illuminate each other

6 立面图
　Elevation

1

3

4

5

6

粤剧艺术博物馆
Cantonese Opera Art Museum

粤剧艺术博物馆是为了保护粤剧文化，传承非物质文化，延续并发展历史城市的社会文化而建设的。

博物馆选址广州传统老城区，项目结合该区域历史街区的改造以及荔枝湾三期改造工程，将整个西关片区的文化串联起来，建成一座展示粤剧艺术和岭南文化的博物馆，一座粤剧艺术文化的大观园。项目用地以荔枝湾涌划分为南北两岸，南岸用地面积11559.5平方米，总建筑面积1.75万平方米，建筑地上3层、地下2层，由主馆、琼花堂、吉庆别馆、銮舆堂、听雨轩、普天乐、广福台、西洋楼、别院声歌等园林建筑围绕中心水池展开，地下一层设陈列展览，地下二层设机动车停车库；北岸用地面积4606平方米，总建筑面积2121.3平方米，建筑地上3层，由三组庭院建筑组成，主要功能为博物馆配套用房。项目总建设投资3.63亿元。

粤剧艺术博物馆在充分发扬现代技术优势之余，立足高远，弘扬地方工艺，尊重传统，与时俱进；项目集传统民间工艺和当代艺术创作等于一体，以岭南风格的园林景观为气息，大量具有地方特色的高标准工艺作品与建筑艺术有机结合，以其中的"三雕两塑"（木雕、砖雕、石雕、灰塑、陶塑）最具亮点。项目集大师精华，赋芸芸匠心，旨在为粤剧艺术的传承、发扬提供空间载体，为时代造园提供研读、探讨的精神场所。

The Cantonese Opera Museum is developed to protect the Cantonese opera, inherit the intangible culture and carry forward the socio-culture of Guangzhou as a historic city.

The Museum is sited in the traditional old urban area of Guangzhou. Combined with the renovation of historical street and Lichi Bay Phase III in the area, the Museum links up all the cultural elements throughout Xiguan Area and create a showplace of Cantonese opera art and Lingnan culture. The site is divided into southern and northern bank by Lichi Bay Riverlet. The southern bank area features a land area of 11,559.5m^2 and a GFA of 17,500m^2 comprising 3 above-grade and 2 basement floors; the above-grade garden buildings, including main venue, pavilion of jade flower, festive hall, practice hall, rain pavilion, pavilion of joy, pavilion of prosperity, western-style pavilion, playing and singing yard, are displayed around a central pool, while B1 is planned for exhibition and B2 for garage purposes. The northern bank area is planned with a land area of 4,606m^2 and a GFA of 2,121.3m^2 comprising three 3-storey courtyard buildings as museum supporting facilities. The total construction investment of the Project is RMB363 million.

While giving full play to modern technologies, the Museum also promotes local techniques and honors tradition. Combining both traditional folk handicraft and modern artistic creation, the design creates a good deal of high-standard handiworks that are organically integrated with architectural art, among which wood carving, tile carving, stone carving, plaster decoration and pottery sculpture are the very highlights. Reflecting the masters' artistic accomplishments and the craftsmen's ingenuity, the Project aims to carry forward Cantonese opera art and provide reference for modern gardening.

项目地点：广州市恩宁路以北，元和街以南，多宝坊以西地段东路
设计时间：2013.12-2014.12
建设时间：2015.1-2016.3
建筑面积：19621m^2
建筑层数：地上3层，地下2层
建筑高度：23.4m
曾获奖项：2017年度广东省优秀工程勘察设计一等奖
2016年度广东省土木建筑学会科学技术二等奖

Location: North of En Ning Lu, south of Yuan He Jie, west of Duo Bao Fang, Guangzhou
Design: 2013.12-2014.12
Construction: 2015.1-2016.3
GFA: 19,621m^2
Floors: 3 aboveground, 2 underground
Height: 23.4m
Award: The First Prize of Excellent Engineering Exploration and Design Award of Guangdong Province, 2017
The Second Prize of Science and Technology Awards of the Civil Engineering and Architectural Society of Guangdong, 2016

1 总平面图
  Site plan

2 错落别致中体会传统建筑的沉静内秀
  Staggered building layout enhances the tranquility and grace of the traditional architecture

3 于无声之处胜有声
  Silence speaks better

4 在园林中幻听戏曲之清律渐行渐近
  The traditional garden offers a perfect background for a fantasy about opera

5 丰富的庭院空间体现传统建筑的构园哲学
  Diversified courtyard spaces embody the traditional garden building concepts

6 市井的音韵在向晚的剪影中回荡
  The museum perfectly fit into the surrounding folk life

2

3

4

5

6

7

8

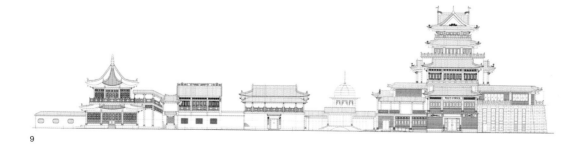
9

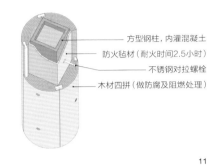

- 方型钢柱,内灌混凝土
- 防火毡材(耐火时间2.5小时)
- 不锈钢对拉螺栓
- 木材四拼(做防腐及阻燃处理)

11

7 走廊围绕中庭空间展开,将人的视线再次引入中庭
  Cloisters around the central courtyard make the latter a visual focus

8 别样的小院情怀
  Courtyard feature

9 立面图
  Elevation

10 传统工艺之匠心独具
   Ingenuous traditional handicraft

11 木包钢柱典型节点
   Typical node of wood-clad steel column

12 民国风的剧场场景复原向粤剧兴盛的年代致敬
   Restored scene of theater during the Republic of China period pays tribute to the then prosperity of Cantonese opera

泰山会展中心
Taishan Convention & Exhibition Center

在泰城时代发展轴线的中部,兴建泰山会展中心,这是一座以会议、展览为主的现代化会展建筑。其中展览部分包括有一个4000平方米的特殊展厅、一个9000平方米综合展厅及两个4500平方米综合展厅,会议中心部分则由一个1250座剧场式会议厅、一个300座报告厅、一个多功能宴会厅、八个小型会议厅组成。另外配套了可停放477辆轿车的地下停车库。

泰安市是中国著名的历史文化名城,如何在建筑形态上表达出泰安丰厚的文化底蕴是本项目在建筑设计专业上的难点和挑战。纵览泰安历史,自尧舜至秦汉,直至明清,泰山都是历代帝王祭天的神山,而鼎正是中华民族祭天最重要的礼器,所以建筑整体外观设计以祭天金鼎为母题,分别用四座鼎作为四个功能区的体量外观,以"盛世金鼎"的整体造型突出泰安市历史文化名城的鲜明形象。外立面则采用一次冲压成形的祥云图形铝合金格栅构件,通过参数化计算形成渐变的排列形式,为四个巨鼎刻画出丰富的表情肌理。

本项目布局紧凑,同时增大了单个展厅面积,提高了展厅使用的灵活性,另外布展方式力求简单直接,方便明了。特别是重点考虑了会展与两会期间共享中庭可分可合的灵活分配,既保证了会议期间的高规格,也保证了会展期间共享空间的完整,解决了会展与会议不同功能使用上的管理问题。本建筑的主要功能布局以展厅为主,都是大空间,层高都在6~8米以上,但是与展厅配套使用的还有很多其他功能的小房间,例如卫生间、洽谈室、茶水间以及为展厅配套的空调机房等。为了协调不同尺度空间的关系,使大空间更纯粹完整,我们在展厅两侧分别设置了两条跨度6米的辅助空间带,把辅助空间放置其中,并且利用夹层设置空调机房,使大小空间的结合更合理有序。

Located in the middle of the urban development axis of Tai'an City, Taishan Convention & Exhibition Center is developed as a modern building mainly for conferences and exhibitions. The exhibition part includes a 4,000m² special exhibition hall, a 9,000m² comprehensive exhibition hall and two 4,500m² comprehensive exhibition halls; while the convention part comprises of a 1,250-seat theatrical conference hall, a 300-seat lecture hall, a multi-functional ballroom, and eight small conference halls. In addition, an underground garage is provided to park 477 cars.

Tai'an is a famous historic city in China. So it is a great challenge for architect to express the profound cultural legacy of the city through building form. From Emperor Yao and Shun in remote ancient times, through the Qin and Han dynasties and till the Ming and Qing dynasties, Mt. Tai had been the sacred mountain for ancient emperors to worship the God of Heaven, while the tripod had been serving as the most important sacrificial vessels in such rituals. Therefore, the overall building form is themed on "heaven-worshiping gold tripod". Four tripod-like volumes are planned to accommodate the four function zones, and the overall form of "the gold tripods in a flourishing age" highlight the distinctive presence of Tai'an as a historic city. The façade features the punch-formed aluminum alloy grille components in the shape of auspicious clouds. These components contribute to varied textures of the four tripods through a gradient matrix deduced through parametric calculation.

With a compact layout, the Project is able to expand the individual exhibition hall, enhance the flexibility in use, and allow for convenient, easy and trouble-free exhibition set-up. In particular, the flexible separation and combination of the shared atrium during the conference period for Tai'an People's Congress and CPPCC and the exhibition period ensures both the high standard of the government conferences and the integral shared space for the exhibition. It is indeed a good management solution for the exhibitions and conferences. On one hand, the main functions in the building are large-space exhibition halls with the floor height of more than 6-8m; on the other hand, there are many other small functional rooms that support the exhibition halls, such as the bathrooms, talk rooms, pantry and AC rooms. To balance the spatial relations between spaces of different scales and ensure the integrity and purity of the large spaces, two auxiliary space belts in span of six meters are provided on both sides of exhibition halls to house the auxiliary spaces. Meanwhile, the AC mechanical rooms are placed on the mezzanine so that the large and small spaces could be combined in a more logic and reasonable manner.

项目地点:山东省泰安市
设计时间:2010.1-2012.3
建设时间:2012.3-2013.4
建筑面积:83317m²
建筑层数:地上2层,地下1层
建筑高度:24m
曾获奖项:2015年度全国优秀工程勘察设计行业奖二等奖
2015年度广东省优秀工程勘察设计工程设计二等奖

Location: Tai'an City, Shandong Province
Design: 2010.1-2012.3
Construction: 2012.3-2013.4
GFA: 83,317m²
Floors: 2 aboveground, 1 underground
Height: 24m
Awards: The Second Prize of National Excellent Engineering Exploration and Design Award, 2015
The Second Prize of Engineering Design under Excellent Engineering Exploration and Design Award of Guangdong Province, 2015

1 总平面图
   Site plan

2 建筑整体外观设计以祭天金鼎为母题
   The overall building form is themed on "heaven-worshiping gold tripod"

3 泰山中轴、盛世金鼎
   Axis of Mountain Tai, gold tripod in the flourishing age

4 祥云图案铝合金镂空外墙,渐变的排列,透光灯光刻画出丰富的表情肌理
   Hollow-out external wall made of aluminum alloy in the shape of auspicious clouds presents diverse fabrics with light on gradient layout

5 祥云图案
　Auspicious cloud pattern

6 主立面实景图
　A view of main facade

7 建筑主入口
　Main entrance of building

8

9

10

11

034 文化、会展建筑 Cultural And Expo Buildings

8 前厅二层实景图
  View of F2 prefunction

9 9000平方米展览空间，满足大规模常规展览要求
  The 9,000m² exhibition space can meet demands of large regular exhibitions

10 从室内看祥云图案铝合金镂空外墙
   The hollow-out external wall made of aluminum ally in the shape of auspicious clouds viewed from interior

11 大天窗提供明亮开阔的空间感受
   Large skylights bring open and bright space

12 1250座剧场式会议厅，满足当地政协会议使用要求
   The 1,250-seat amphitheatrical conference hall can meet demands of Tai'an CPPCC meetings

13 300座报告厅室内实景图
   Interior real scene of a 300-seat lecture hall

广东（潭洲）国际会展中心首期工程
Guangdong (Tanzhou) International Convention and Exhibition Center (Phase I)

广东（潭洲）国际会展中心首期工程总建筑面积约为11.55万平方米，定位为集展览中心、会议中心、商务中心和接待中心于一体的复合功能核心场馆，主要建筑包含5个展馆、1个登陆厅和会议中心、相关配套设施。它是顺德提升城市形象，完善城市功能，打造城市名片的标志之一。

建筑造型以折纸及剪纸艺术为设计灵感，通过翻折的手法，形成高低起伏的建筑形象，暗合了建筑的不同功能，形象自然而有力量，建筑以充满力道的折线为主，连续的折线屋面将展馆覆盖，形成连续界面。

展厅为大空间设计，按照德国汉诺威会展公司要求建造。展馆的布局强调实用性和便捷性，展馆定位为重型机械工业展厅，展厅地面荷载要求为100千牛／平方米，大空间结构内要求设置大量吊挂点以满足布展要求。根据建设工期，选用了技术成熟、施工简便的倒三角桁架受力体系；桁架之间采用钢梁进行连接，既保证桁架平面外的稳定，同时也满足了后期使用满布预留吊点要求。梁和桁架的节点采用螺栓连接，减少恶劣天气对施工的影响，缩短了施工工期，符合装配式建筑的发展方向。

Guangdong (Tanzhou) International Convention and Exhibition Center (Phase I) with a GFA of about 115,500m^2 is defined as a mix-use major venue integrating the exhibition center, convention center, business center and reception center and including 5 exhibition halls, 1 concourse, convention center and supporting facilities. This representative project is a part of Shunde's efforts to upgrade the urban image, improve the urban function and build the urban icon.

Inspired by the art of paper folding and cutting, the building takes the undulating shape by folding, suggesting different building functions with natural and powerful presence. Dominated by the sweeping fold lines, the building sees continuous surface generated by the extended knee roofs of exhibition halls.

The large space exhibition hall is constructed by following the requirements of Hannover Fairs International GmbH. The exhibition hall is designed with pragmatic and convenient layout, the floor load of 100 kN/m^2 and large number of hoisting points for the setting-up of heavy machinery industry exhibition. In view of the construction period, inverted triangle truss load-bearing system is selected, which is technically mature and easy for construction. The truss connected by steel beam ensures the stability beyond the truss plane and the hoisting points reserved for later operation. The bolt connection between the beam and truss node minimizes the impact of bad weather on the construction and shortens the construction period, which is in line with the development trend of prefabricated building.

2

1

项目地点：广东省佛山市顺德区
设计时间：2016.1
建设时间：2016.2-2016.9
建筑面积：115549.47m^2
建筑层数：地上3层，地下1层
建筑高度：最高点25.721m
合作单位：同济大学建筑设计研究院（集团）有限公司

Location: Shunde District, Foshan City, Guangdong Province
Design: 2016.1
Construction: 2016.2-2016.9
GFA: 115,549.47m^2
Floors: 3 aboveground, 1 underground
Height: 25.721m at the highest point
Partner: Tongji Architectural Design (Group) Co., Ltd.

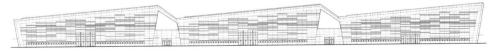

1. 总平面图
   Site plan

2. 通过翻折的手法，创造出简洁的折线立面形象
   Concise folded façade through folding approach

3. 建筑造型简洁，形象自然而有力量
   Concise, natural and powerful building presence

4. 渐变的遮阳百叶，有效地实现绿色节能，同时丰富立面效果
   Gradient sunshade lamellas allow for effective energy saving and diversified façade effect

5. 通过错落的穿孔金属板，使规整的建筑立面充满动感
   The staggered perforated metallic sheets make the regular building façade dynamic

6. 立面图
   Elevation

7

8

9

10

7 展厅空间宽阔方正，为展览提供很好的展示场所
The spacious and square exhibition hall offers ideal exhibition space

8 各种设备整齐有序地布置在展厅的天地墙上，极大地满足了展览的使用需求
Various equipment orderly distributed on the floor, ceiling and walls of the exhibition hall greatly satisfies the exhibition demands

9 明亮的连接厅为参观者提供较好的休息空间和共享空间，同时也为展厅提供了较好的功能补充
The daylit connecting hall offers pleasant space for visitors to rest and communicate, meanwhile complements the functionality of the exhibition halls

10 宽敞通透的连廊为参观者提供较好的步行空间
Spacious and transparent connecting bridge offers nice pedestrian space for the visitors

科技教育建筑

科教建筑一般包括学校建筑和科研实验建筑,与一般公共建筑相比,专业性更强,功能要求更严谨,甚至具有工业建筑的某些特征。在65年的科教建筑设计实践中,GDAD的建筑师通过对其专业性的要求进行深入理解和研究,针对不同的个案,运用合理的建筑逻辑关系、空间秩序、建筑形式,营造出与特定需求相匹配的工作、学习环境,创作出一大批在工艺、管理模式与建筑创造的高度统一的优秀作品。如珠海翔翼保税区项目是目前国内仅有的三家飞行员培训基地之一,建筑师通过对特殊的工艺(流线)要求、管理的要求和空间形态要求的分析总结,采用符合分期建设的有机生长模式,按照利于综合管理的原则进行放射性生产经营区布局,采用简洁现代的造型手法及有机的建筑组合方式,打造出工业与民用建筑功能和形式上的完美结合产物。

对于传统的学校尤其是历史名校的建筑设计中,除了要保证教与学的基本使用功能要求外,GDAD的建筑师还十分注重学校自身的文化氛围和历史文脉的研究,更十分注重体现与时俱进的时代感。如北京师范大学教学办公楼的设计,建筑师采用严谨均衡的布局,并与已有的图书馆共同组成教学主轴建筑群。建筑空间和建筑形象的塑造层次分明、方正大气,较好地诠释了学校严谨的治学精神和自信的京师风范。

随着社会的进步、科技的发展,教学科研的对象、模式、手段也产生日新月异地创新变化,使用者的行为习惯和精神追求也越来越需要重视和研究。横琴创意谷(横琴澳门青年创业谷)项目一改科研建筑常见的方正刻板布局和形象,采用创享生态圈的设计概念,通过折线形体和体块搭接、退台和架空的手法,形成灵动多变、内外相融的空间效果。打造出富有魅力和活力、富有弹性、满足发展需要的建筑空间环境。

地域性是当代建筑设计的基本原则之一,这不仅体现当代人回归自然的追求,更是最大限度利用大自然,展示当地特色和建筑个性的一种手段。广州气象雷达站、华润大学、广东青年干部学院钟落潭新校区项目,在满足功能要求的同时,因地制宜、因山就势,建筑与其特殊的环境地理位置紧密结合。在设计中结合岭南气候特点,组织庭院空间,体现出GDAD作品一贯尊重环境、强调建筑地域性的设计理念。

城市新陈代谢的进程,投入悠长的历史中,只是瞬息。建筑的可持续发展不仅满足人与自然环境和谐发展的要求,也符合我国全面协调发展战略,记住承载着某一时代集体记忆的建筑,更是延续城市生命、重生城市文化的重要手段。ADG·机场设计研究院办公楼(信义会馆)由一座红砖厂房与一栋新建的两层钢结构建筑组成,建筑师通过娴熟的空间组织手法和对不同材料的选择应用,使新旧建筑形态产生碰撞与融合,钢结构、玻璃、红砖等现代与传统材料调和结合,既体现建筑的现代性,也表达了对传统的尊重,旧建筑活化出新的生命。而广州正佳极地海洋世界是国内首个在原有商场基础上改造的海洋馆项目,拥有全球最大的曲面亚克力水族展示面。改造设计采用了框架结构体系改造为框架-剪力墙结构体系、增加阻尼器与耗能斜撑、提高建筑物地震下延性等多项创新的结构加固和改造技术,满足了海洋馆丰富的空间形态、严格的营运管理和复杂的设备要求,体现了GDAD人勇于面对挑战,善于解决复杂技术难题的高超水平。该项目也成为商业综合体再开发利用升级的成功案例。

S & T and Academic Buildings

S & T and academic buildings generally include school buildings and scientific research lab buildings. Compared with general public buildings, they require more specialized design and impose more stringent requirements on functions, even to a degree of bearing some characteristics of industrial buildings. In the past 65 years of design practices in this field, GDAD's architects have deepened their understanding and research of the specialized requirements, and, on a case by case basis, carefully coordinated the architectural logic, spatial order, and forms of buildings to create the environment for work and study meeting specific needs. So far we have created a large number of exceptional works that achieve a high degree of harmony in the technology, management model and architectural design. One example is the project of Xiangyi Bonded Area, Zhuhai, one of the only three domestic pilot training bases in China. After analyzing the requirements for special process (circulation), management and spatial form, we adopted an organic growth mode to accommodate phased development, arranged the production and operation areas in a radial form to facilitate the comprehensive management, and realized the perfect combination of industrial and civil buildings in both functionality and form through concise modern building shape and organic building composition.

In the architectural design of universities, in particular those prestigious universities, in addition to the basic functions of teaching and learning, we also emphasize the universities' culture and history, as well as the spirit of the times. For example, in designing the Teaching & Office Building of Beijing Normal University, we adopted a rigorous and balanced layout to create the complex on the main teaching axis together with the existing library. With coherent and generous building spaces and appearance, the building properly interpret the university's rigorous scholarship and self-confidence as a normal university in the capital city of China.

The social and technological development has driven the constant innovation and changes to the subjects, modes and means of teaching and research, requiring more attention and researches on the users' behavioral habits and spiritual pursuit. In designing the project of Innovalley Hengqin, we abandoned the stereotype of rigid layout and appearance commonly found with S & T buildings, and instead adopted the concept of innovation and sharing eco-circle. This way, we were able to, through various approaches like zigzagged building form, building block splicing, setback and open-up floors, create flexible and varied spaces to integrate the interior with the exterior and present a creative, lively, and resilient built environment for future development.

The feature of place is one of the basic principles of contemporary architectural design. It not only reflects people's desire of returning to nature, but also showcases the local characteristics and architectural identity to the maximum through the nature. Guangzhou Weather Radar Station, China Resources University and Zhongluotan New Campus, Guangdong Youth Leader College projects are vivid examples of the principle. While meeting the functional requirements, the buildings of these projects are designed with their specific locations and mountainous setting taken into consideration. Moreover, the climatic features of Lingnan region are also considered in organizing the courtyard spaces, reflecting our consistent architectural design ideas of respecting the context and the features of place.

The process of urban evolution is merely a transient moment in the long history. The sustainable building development is not only necessary for human-nature harmony and development, but also consistent with the comprehensive and coordinated development strategy of China. Buildings that carry the collective memory of a generation, is an important means to renew the urban life and revive the urban culture. ADG Office Building (Xinyi Place) consists of a red-brick plant and a new two-story steel structure. The ingenious spatial organization and material application enable the old building forms to collide and integrate with the new, and the modern materials of steel structure and glass to blend with the old red bricks. This reflects the modernity of the buildings and the respect for traditions, injecting vitality into the old buildings. The Grandview Aquarium is the first shopping-mall-turned aquarium in China, featuring the world's largest curved acrylic display surface. For this project, various innovative structural reinforcement and transformation technologies are employed, such as transforming the original frame structure into the frame-shear wall structure, adding damper and energy dissipation brace, and improving the seismic ductility of building structure. Such measures properly accommodate the aquarium's demands for varied spatial forms, stringent operation and management demand and sophisticated MEP system, and showcase our professionalism and ingenuity in meeting challenges and tackle complex technical problems. The project has set up a good example for reuse and upgrading of commercial complex.

北京师范大学教学办公楼
Teaching & Office Building of Beijing Normal University

北京师范大学校园内，南北主轴线上，新建北京师范大学教学主楼，正对学校南门。工程用地南北约400米，东西约102米。轴线上已有5层图书馆一座。新建建筑群包括新主楼（南楼）和信息中心（北楼），分别位于图书馆的南侧和北侧，与已有的图书馆，共同组成"教学主轴"建筑群。它将作为整个学校的新主楼，同时也是标志性建筑。

教学办公楼总建筑面积为73,668平方米，其中地上建筑面积为56,795平方米，地下室面积16,873平方米。建筑地上23层，地下2层，总建筑高度90.55米。地下2层主要为藏书库、汽车库、设备用房，地上一至八层为图书馆用房；地上九至二十三层主要为该校文科各系的办公、学术用房等。新办公楼为8层，主要为行政办公用房，首层有财务室等，设1层地下室，内有设备用房及校史资料陈列室等。

新主楼面向校前广场，是校园南门的对景，也是校园南北主轴线上空间序列的第一个高潮。高层部分有大于1/4周长直接落地，消防车道沿建筑物周边布置。东西两侧连廊首层高度为5.4米，消防车可以进入。本工程与周围建筑间距大于13米，基本保留了原有道路系统。

Located on the north-south axis of the campus, the newly-built main Teaching & Office Building of Beijing Normal University faces the South Gate. The project site measures 400m from north to south and 102m from west to east. Together with the five-floor existing library on the axis, the new buildings, including a new main Teaching & Office Building (South Building) and an Information Centre (North Building) that sit to the south and north of the library respectively, form the building cluster on the Teaching Axis. As the new main building on the campus, it will also be the landmark of the University.

The GFA of the Teaching & Office Building reaches 73,668m^2, including 56,795m^2 and 16,873m^2 respectively for the above ground development and the basement. The building is 90.55m high with 23 above ground floors and two basement floors. The two basement floors accommodate book storage rooms, garages and mechanical rooms. The first to the eighth floors are the library and the ninth to the twenty-third floors are the offices and academic rooms for the schools of liberal arts. The new Administrative Office Building has eight floors that accommodate mainly the administrative offices with the Finance Office on the first floor. This Building has one basement floor for mechanical rooms and the university history exhibition rooms.

The new building facing the square in front of the University becomes an opposite view of the South Gate and the first climax in the spatial sequence along the north-south axis. For the high-rise part, over 1/4 of the external walls land to the ground directly, with the fire lanes surrounding the Building. The first floor of the corridor on the east and west is 5.4m high, allowing the fire trucks to drive in directly. The interval between the project and the surrounding buildings exceeds 13m, and the existing road systems are basically retained.

项目地点：北京市北京师范大学
设计时间：2000
建设时间：2010
建筑面积：73668m^2
建筑层数：23层
建筑高度：90.55m
曾获奖项：2011年度广东省优秀工程勘察设计奖二等奖
2010年北京市建筑长城杯金质奖
2007年北京市结构长城杯金质奖
2005年度广东省第十二次优秀设计奖一等奖

Location: Beijing Normal University, Beijing
Design: 2000
Construction: 2010
GFA: 73,668m^2
Floors: 23
Height: 90.55m
Awards: The Second Prize of Excellent Engineering Exploration and Design Award of Guangdong Province, 2011
                Gold Award of Beijing Architecture Great Wall Cup, 2010
                Gold Award of Beijing Structure Great Wall Cup, 2007
                The First Prize of the Twelfth Excellent Design Award of Guangdong Province, 2005

1

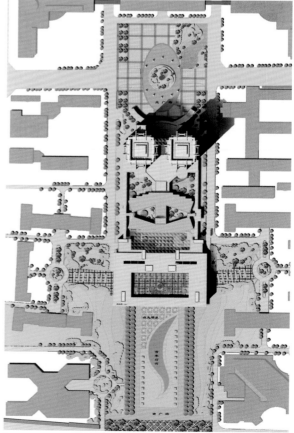

2

1 主入口气势恢宏，方正严谨，形象极佳
  The main entrance is impressive with a magnificent, square and meticulous appearance

2 总平面图
  Site plan

3 整体形象前低后高，层次分明，功能合理，轴线清晰
  The building cluster, with building heights increasing from the front to the back, presents a distinct architectural hierarchy, reasonable function layout and a clearly-defined axis

4 高层办公塔楼
  High-rise office tower

S & T and Academic Buildings 科技教育建筑　043

珠海翔翼保税区项目
Project in Zhuhai Xiangyi Bonded Area

珠海翔翼保税区项目是目前国内仅有的三家飞行员培训基地之一。项目总平面、平面功能、工艺、流线等设计合理，整体造型新颖、美观，与周边环境协调统一。在工业及工艺建筑与民用建筑完美结合方面进行了创新和研究，从建筑设计上对飞行员训练管理模式与建筑平面功能的高度统一方面进行了创新和研究。建成后，得到了业主、施工单位等多方面的高度评价，成为国内同类型项目的样板及参观基地。

珠海翔翼保税区项目分为生产经营区和飞行员休息区。生产经营区总体设计为可以满足40个模拟机教学及其他配套的设施，飞行员休息区按650个床位的四星级酒店标准进行设计。由于飞行员培训功能的特殊性，该项目既具有工业建筑的特点，又具有民用建筑的特点，因此在设计中充分考虑项目性质特点进行空间及造型设计。由于模拟机非常昂贵，因此在设计中充分考虑模拟机大厅的组合设计、工艺流程、设备布置及综合管理。

项目采用符合分期建设的有机生长模式，充分考虑分期建设的可实施性及各阶段的建设完整性；按照利于综合管理的原则进行放射性生产经营区布局，提高整个生产经营区的管理效率；设计园林式景观布局的飞行员休息区，为飞行员提供了休闲的居住环境；采用简洁现代的造型手法及有机的建筑组合方式，打造工业与民用建筑功能和形式上的完美结合产物。

As one of the three pilot training bases in China, the Project in Zhuhai Xiangyi Bonded Area features well-thought site plan, functional layout, processes and circulations. The novel and aesthetical building form is well embedded into the context. For this project, innovation and researches are made as for how to perfectly integrate the industrial and technical building with the civil building and how to unify the pilot training and management function with the functional layout. The project is highly appraised by the client and contractors as a frequently visited benchmark of its kind in China. .

The project comprises of the production and operation area and the pilot accommodation area. The former is designed to accommodate 40 training simulators and other supporting facilities, while the latter is designed according to the standards of a 650-bed four-star hotel. In light of the particularity of pilot training function, the project incorporates the features of both industrial and civil buildings, which has to be fully considered in the design of space and building form. As the simulators are very expensive, the composition, technical process, equipment layout and management are also carefully considered in the design of the simulator hall.

The Project employs an organic growth mode that well fits the phased development to fully accommodate the viability of the phased construction and the integrity of each phase. Radial layout in the Production and Operation Zone follows the principles of the comprehensive management and enhance the management efficiency. The garden-like setting of the Pilot accommodation Area offers a relaxing living environment for the pilots. The concise and modern architectural form and the organic building composition contribute to the perfect combination of industrial buildings with civil buildings, both functionally and formally.

项目地点：珠海保税区
设计时间：2006.3-2006.8
建设时间：2006.9-2009.12
建筑面积：101819.3m²
建筑高度：19.7m
建筑层数：4层
曾获奖项：2013年全国优秀工程勘察设计行业奖二等奖
2011年度广东省优秀工程勘察设计奖二等奖

Location: Zhuhai Bonded Area
Design: 2006.3-2006.8
Construction: 2006.9-2009.12
GFA: 101,819.3m²
Height: 19.7m
Floors: 4
Awards: The Second Prize of National Excellent Engineering Exploration and Design Award, 2013
The Second Prize of Excellent Engineering Design Award of Guangdong Province, 2011

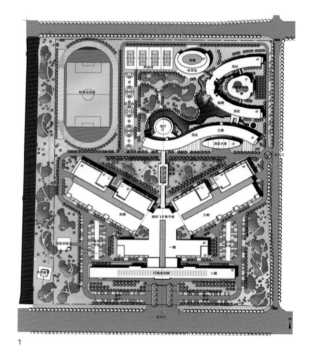

1 总平面图
  Site plan

2 采用简洁现代的造型手法及有机的建筑组合方式，创造简洁、明快、现代的建筑群体
  Concise and modern architectural form and the organic building composition jointly create a simple yet lively and modern complex

3 以现代的手法重新演绎点、线、面的关系，利用体块的穿插和组合形成丰富的造型
  Modern approaches are employed to reinterpret the relation between the point, linear and aerial form. The interwoven and combination of building blocks jointly create diverse building form

4 行政办公楼采用简洁、以横向线条为主的铝板及玻璃幕墙，入口采用大片玻璃幕墙，强烈的虚实对比丰富了建筑的外立面
The administration office building features concise aluminum panels and glass curtain walls mainly in horizontal lines, while the entrance is provided extensive glass curtain walls. This sharp contrast between solidness and void diversifies the building facade

5 代表企业形象的"翔翼蓝"与现代灰相结合，打造简洁、明快的三层中空入口大堂
The iconic "Xiangyi Blue" is combined with the modern grey to present a concise, lively and light-flooded entrance void in three-floor height

6 位于三条放射形生产经营区交汇处的中央管理室，有效提高生产经营管理效率
The central management room at the junction of three radial production and operation strips effectively enhances the operation and management efficiency

广州气象雷达站
Guangzhou Weather Radar Station

广州气象雷达站是为安装当时世界上最先进和最精确的天气雷达系统CINRAD而兴建，位于广州番禺大镇岗，其山顶是方圆数公里的最高点。在满足气象监测功能的前提下，气象站设计尝试将气象建筑与其特殊的环境地理位置相结合，使其成为该区域的景观建筑，并努力成为科普、旅游和气象系统会议的良好场所。从建成以来的使用情况看，这些设想已全部或大部分地得以实现。

雷达站位于山顶，建筑范围地势高差达7米。该项目分四期工程。一期办公楼采用三角形切去三个角而得到的不等边六边形平面形状，使建筑物具有较好的稳定性，满足其上部雷达天线转动时对支座稳定性的技术要求。又因办公楼位于山顶制高点，这种平面形状还有利于建筑物形象对各个方向的视觉一致性，更好地令其独特的体型成为建筑群的标志。二期、四期员工宿舍和三期展厅、会议室均依山势迭级而建，并在外形上保持了办公楼多边形的特点。三个展厅套叠而建，活跃了建筑群的氛围。首期与二、三期之间的露天的三向斜交井格梁，与各级内庭一道，使建筑物内外空间更具流动性。休闲泳池及瀑布平台令建筑群与室外环境结合得更为完美。这些处理手法是对岭南山地建筑创作的新的尝试。

建筑物外墙采用红砂岩面砖为主，与山体土质、颜色相呼应，坡屋顶采用"薄石片蓝灰瓦"贴面，配合建筑造型，更好地体现了山地建筑的特点。这些处理使得建筑物成为山体的合理延伸，并成为该山的良好点缀，球形的白色雷达天线罩对整个建筑群起到了点睛的作用。项目建成后，以其别具一格的形象受到国内气象系统的关注，并成为广州南郊一科学景点。

Standing at the top of the Dazhengang Hill in Panyu District, the highest point in several kilometers around, Guangzhou Weather Radar Station was built to house CINRAD Radar, the then most sophisticated and precise weather radar system in the world. While meeting the demands for meteorological monitoring, the Station was also envisioned to combine the meteorological functions with its unique location to create a view for the area and an ideal place for science communication, sightseeing and meteorologists' meetings. So far these visions have been proven partially or wholly realized.

The Station stands at the hilltop, with a 7-meter height difference between the site and the surrounding areas. The project comprises of four phases. The Phase I office building features a planar geometry of non-equilateral hexagon which is generated by cutting off three corners of a triangle. Such geometry makes the building more stable, satisfying the technical requirements on the stability of the bearings when upper radar antenna rotate. Moreover, it helps create a consistent building image when viewed from all directions thanks to the building's hilltop location, and make this unique form an icon of the complex. The Phase II & IV staff dormitory and Phase III exhibition halls and the meeting rooms are staggered along the slope, sharing the polygonal feature of the office building. Three exhibition halls interconnect and stagger with each other to make the whole complex alive. Coupled with the various levels of inner courtyards, the open-air three-way diagonal grid beams between Phase I, II and III enhance the fluidness of the interior and exterior spaces. The leisure swimming pool and the waterfall terraces perfectly combine the building complex with the outdoor environment. The aforesaid approaches are some experiments with the architecture design in the mountainous Lingnan area.

The facade cladding of red sandstone tiles echo to the texture and color of the earth on the hill. The sloping roofs are clad with thin stone tiles in blue grey to match the architectural form and demonstrate the features of buildings in mountainous region. This way the buildings become natural extension and highlight of the hill, and the spheroidal white radar dome a crowning touch of the complex. With its unique building form, the completed complex attracts attention from the Chinese meteorological circle and is much welcomed as a science destination in the southern suburbs of Guangzhou.

项目地点：广州番禺
设计时间：1998.10-1999.7
建设时间：2000.6
建筑面积：4910m²
建筑层数：7层一幢、4层两幢、1层一幢
建筑高度：27.9m
曾获奖项：2001年获广东省第十次优秀工程设计二等奖
2001年获广东省注册建筑师协会首届优秀建筑创作奖

Location: Panyu, Guangzhou
Design: 1998.10-1999.7
Construction: 2000.6
GFA: 4,910m²
Floors: one building with 7 floors, two with 4 floors, and one with 1 floor
Height: 27.9m
Awards: The Second Prize of the Tenth Excellent Engineering Design Award of Guangdong Province, 2001
The First Excellent Architecture Creation Award by Guangdong Chapter of Association of Chinese Registered Architects, 2001

1 层叠错落的建筑犹如自然山体的一部分
  The staggered and cascading buildings appear to be a natural part of the hill

2 最初的设计草图中已经表达了建筑与山体融为一体的建筑意向
  The original design sketches already show the intent to integrate the buildings with the hill

3 总平面图
  Site plan

1

2

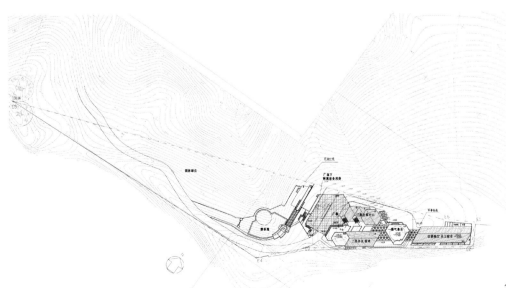

3

4

5

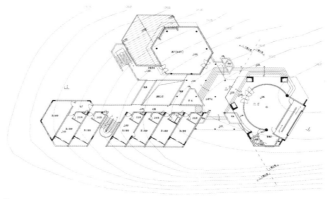

6

4 矗立在山顶的雷达站
　Radar station at hilltop

5 建成的实景体现了最初构思的场景
　The built scene shows the initial concept

6 二、三期三层平面及一期首层平面
　Phase II & III F3 floor plan and Phase I F1 floor plan

7 坡屋顶、平台、走廊相互交错既富有趣味性，同时又使得视觉空间更为立体
The interconnected sloping roofs, terraces and corridors jointly present a more interesting and diverse view

8 结构网格再现了平面构成逻辑
The structural grid recreates the logics of planar composition

9 楼梯和两侧墙体脱开，犹如嵌在石缝中
Stairs are disconnected from walls on both sides, as if being embedded into the rocks

10 利用山体的高差形成自然庭院
Courtyards are naturally formed by taking advantage of the height difference of the hill

ADG·机场设计研究院办公楼
ADG Office Building

信义会馆是一个由工业厂区改造而成的办公园区。它坐落于广州芳村白鹅潭畔，面朝珠江。茂密的老榕树、青砖路与红砖厂房互相映衬，形成了园区内独特的环境氛围。因此，一些创意产业，如广告公司、摄影工作室、画廊、设计公司等纷纷选择这里作为办公场地。

ADG·机场所的办公楼由一座红砖厂房与一栋新建的两层钢结构建筑组成。信义会馆的业主为我们提供了一个设计自己办公空间的机会，建筑及装修考虑了团队的现实需求与发展规划。老厂房内部的大空间完全被保留下来，为设计师提供了开放、宽敞的办公场所。厂房瓦顶下方满吊彩钢夹芯屋面板，改善了瓦屋顶隔热、漏雨、瓦片跌落等问题。新建的钢结构部分则提供独立办公室、会议室等小空间。大面积的落地玻璃窗把园区优美的景色全部纳入内部的办公空间中。办公楼大堂通高两层，从入口处可以透过二层的玻璃走廊看到老厂房的坡屋面，形成独具特色的入口景观。精细的钢结构建筑与粗犷的红砖厂房巧妙地结合在一起，内部空间功能划分清晰，使用高效，从建筑形式、空间、色彩、材质等多方面体现了现代与传统的碰撞和融合。

这座办公楼是一个旧建筑活化使用的好案例。同时，它也承载着ADG·机场所团队"好设计、用心做"的经营理念，是团队的一张重要名片。

Xinyi Place is an office park transformed from an industrial zone that is located at White Swan Waterfront, Fangcun, Guangzhou, facing the Pearl River. Here, leafy old banyans, grey brick roads and red brick plant buildings set one another off nicely to form a unique atmosphere. Dozens of advertising companies, photography studios, art galleries and design firms from the creative industry have flocked to set up their offices.

ADG's office building is composed of a red brick plant building and a new two-storey steel structure. The owner of Xinyi Place offered us an opportunity to design our very own office space, so our practical demands and future development plan were fully considered in the architectural and fit-out design. The generous space inside the old plant building is fully retained to provide an open and spacious office for the designers. Under the tile roof of the plant building, color steel sandwich roof panels are suspended to cover the whole interior space, addressing problems like thermal insulation, rain leakage and tile drops. The new steel structure houses the individual offices, conference rooms, and other small closed spaces, while the full-height glass windows bring the pleasant exterior views into the interior. At the entrance of the two-floor lobby, the sloping roof of the old plant building is visible through the F2 glass corridor, presenting a unique view at the entrance. The fine steel structure and the rough red brick plant building are perfectly combined to form an interesting contrast, while the interior spaces and functions are clearly defined for efficient uses. Such collision and integration of modernity and tradition are reflected in many ways from architectural form to building space, tone and materiality.

This office building sets a good example of reinvigorating and reusing the old buildings. It is an icon that embodies and represents ADG's business philosophy, i.e. design with heart and soul.

项目地点：广州芳村信义会馆
设计时间：2008-2009
建设时间：2009
建筑面积：1273m²
建筑层数：2层
建筑高度：9m

Location: Xinyi Place, Fangcun, Guangzhou
Design: 2008-2009
Construction: 2009
GFA: 1,273m²
Floors: 2
Height: 9m

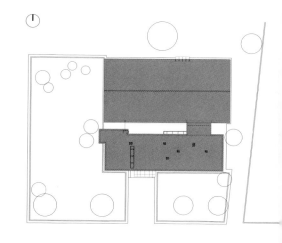

1

1 总平面图
　 Site plan

2 钢结构、玻璃、红砖等现代与传统材料调和结合，既体现建筑的现代性，也表达了对传统的尊重
　 Steel structure, glass and the red bricks jointly present the sense of modernity while showing respect for the tradition

3

4

6

7

3　改造前的信义会馆办公楼，前身是一座红砖厂房
Predecessor of ADG Office Building was a red brick plant building

4　改造后的背立面，在原有建筑基础上增加通透玻璃
Renovated back elevation is added with transparent glazing

5　从入口门厅可以看到二层的玻璃廊道、花园的绿植及厂房坡屋面，形成独具特色的入口景观
F2 glass corridor, greenery in the garden and the sloped roof of the plant building are visible from the entrance lobby, offering an unique view at the entrance

6　新增的钢结构正立面
The front elevation of the added steel structure

7　新与旧并存融合的建筑立面
Building elevation integrates the old with the new

8　位于新旧建筑之间的二层露台花园，形成舒适的休憩空间
F2 Terrace garden between the new and the old building offers a pleasant leisure space

华润大学
China Resources University

华润集团为全面实践华润品牌战略目标，系统提升组织能力，推动集团可持续发展，新建华润大学，作为重要人才的培训基地。坐落于深圳东海岸的惠州霞涌小径湾，华润大学依山而建，傍海就河，微风徐徐，林荫茂盛。矗立在山顶，看星光大道直通天边；扶栏眺远，看夕阳落入云海深处。

华润大学以山顶星光大道为景观主轴，以轴线南端处主体培训楼作为主体形象，两侧有对称展开的教学楼区建筑群落；在主体教学楼区东南侧为综合馆区，布置有会议、运动、图书馆及酒堡等功能；在主体教学楼区东侧顺坡而下建有两栋招待所，为招待所区。三大区域以南北向轴线为核心，顺应山势，设计阶梯山道步行系统，将每个平台有机且富有趣味地组织起来，形成一个完整生动的建筑群落。

"大学之道，在明明德，在亲民，在止于至善。"其之本在于修心。建筑艺术不仅注重空间与形式的创作，也涵盖了细部设计与建造工艺的营造，华润大学营造之道在于"匠心"。其一，师法自然，"取胜""形势"。其二，红砖清混，至美至善。其三，力学之美，协调之美。还建筑以质感，给大学静谧以优雅，妙不可言；大学至臻，精心至简；星光大道树影婆娑，笑聚八方之才。

China Resources University (the "Project") serves as the talent training base of China Resources Group to comprehensively practice its brand strategic objective, systematically improve the organization ability and promote its sustainable development. The project is located in Xiaojing Bay, Xiachong, Huizhou to the east of Shenzhen, lying against the hills by the sea and the river and enjoying gentle breeze and lush trees. Standing on the mountain summit, one may have a spectacular view of the Star Avenue that leads directly to the horizon, or lean against the handrail overlooking the sun setting deep into the clouds.

With the Star Avenue as the main landscape axis, the Project takes the main training building at the southern end of the Axis for project presence. The teaching buildings are symmetrically extended along both sides to form the teaching building zone. On the southeast of the main teaching building zone is the comprehensive venue zone including such functions as meeting rooms, sports facilities, library and cellar, while on the east down the slope two guesthouses are provided down the slope, forming the guesthouse zone. Centering on south-north axis, the three major zones follow the slope with stepped hill footpath system to connect all platforms in an organic and interesting way for a complete and dramatic building cluster.

"The aim of Great Learning lies in the advocacy of brilliant virtues, the remolding of people, and the pursuit of ultimate goodness. Its essence is to cultivate one's mind". Architecture not only stresses the creation of space and form, but also covers the development of detail design and construction technologies. The design approach of the Project emphasizes ingenuity. Specifically, it finds inspiration from nature and excels in the terrain; red bricks and fair-faced concrete are blended beautifully; the aesthetics of mechanics and harmony are also highlighted. All these give buildings an impressive texture and create a tranquil and elegant environment. The University strives for excellence, starting from the simplicities. Lined with swaying trees, the Star Avenue shows a welcome gesture to talents from all over the country.

项目地点：广东省惠州市霞涌小径湾
设计时间：2013.3-2014.9
建设时间：2013.6-2016.7
建筑面积：53585.37m²
建筑层数：5层
建筑高度：23.9m
合作单位：英国福斯特及合伙人事务所
曾获奖项：第三届深圳市建筑工程施工图编制质量金奖

Location: Xiaojing Bay, Xiapu, Huizhou City, Guangdong Province
Design: 2013.3-2014.9
Construction: 2013.6-2016.7
GFA: 53,585.37m²
Floors: 5
Height: 23.9m
Partner: Foster + Partners, U.K.
Award: The Gold Award of the Third Shenzhen Project Construction Drawing Quality

1　上帝视角总平面
　　Master plan from the God's perspective

2　鸟瞰建筑群落在山间散落（南望向北）
　　Buildings scattered in the hills from the bird's eye view (look northwards from south)

3　鸟瞰建筑群落往海边与山间延伸（北望向南）
　　Buildings extending to the seaside and hills (look southwards from north)

4　小径湾大鸟瞰望向海边（西北侧望向东南）
　　Bird's eye view of Xiaojing Bay till the seaside (view from northwest to southeast)

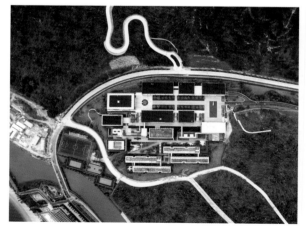

1

S & T and Academic Buildings 科技教育建筑 055

5　星光大道主楼夜景图
　　Night view of the main building along Star Avenue

6　红砖墙廊道
　　Corridor enclosed by red brick walls

7　9栋图书馆单侧悬空楼梯
　　Single-side suspended stairs of Building 9 (the Library)

正佳海洋世界生物馆
Grandview Aquarium

在广州市第一商圈——天河商圈的核心地段，天河路南侧，广州市中轴线东侧的正佳商业广场，是全亚洲单体规模最大的商业综合体，年均销售额及人流量全国第一。为提升商业氛围，增加独特体验，大力吸引顾客，在正佳商业广场里新建海洋世界生物馆。

广州正佳极地海洋世界是国内首个在原有商场基础上改造的海洋馆项目，拥有全球最大的曲面亚克力水族展示面，宽度达到40.8米。设于广州天河正佳商业广场内部西南角二至四层，该位置原为友谊商店，建设单位拟将其改造为正佳海洋馆。涉及改造范围面积为18790平方米。海洋馆外的商场区域不做改动。总建筑面积：18790平方米（改造范围），建筑层数：7层，其中改造楼层为二至四层，原结构形式：框架结构，改造后结构形式：框架-剪力墙结构。

海洋馆平面按使用功能划分为4种空间区域，分别为：游客可达区域、展缸区域、维生设备区域、后勤及疏散区域。各个功能区之间相互依存、相互制约，需要不断反复推敲才能协调好建筑、结构、设备、室内装修、维生系统、亚克力展示板安装、声光设备、运营管理等各个专业的需求。除了复杂的平面布局及流线组织外，海洋馆改造工程的另一难点在于空间剖面设计。与新建海洋馆不同，原商场层高非常有限（层高5米），而且由于要对原有结构进行加固，海洋馆区域范围内需增加一层400~600毫米厚的加固楼板，使空间高度更为局促。结构改造设计充分考虑了海洋馆大荷载的影响，采用了截柱增加转换梁加固、主缸底板区新型梁柱加固节点、增加钢板剪力墙，将原框架结构体系改造为框架-剪力墙结构体系、增加阻尼器与耗能斜撑，提高建筑物地震下延性等多项结构新技术确保建筑物安全，是原有商业综合体再开发利用升级的成功案例。

项目地点：广州市天河区天河路228号
设计时间：2014.5
建设时间：2016.2
建筑面积：18790m²
建筑层数：地上2-4层
合作单位：澳洲PTA建筑设计有限公司
获得奖项：2016年度广东省土木建筑学会科学技术奖一等奖
2017年度广东省优秀工程勘察设计二等奖

Location: 228 Tian He Lu, Tianhe District, Guangzhou
Design: 2014.5
Construction: 2016.2
GFA: 18,790m²
Floors: 2-4 aboveground
Partner: Australia PT Design Consultants Limited
Award: The First Prize of Science and Technology Awards of the Civil Engineering and Architectural Society of Guangdong Province, 2016
The Second Prize of Excellent Engineering Exploration and Design Award of Guangdong Province, 2017

Grandview Aquarium is located in the core of Tianhe Commercial Circle, the No. 1 commercial circle of Guangzhou, on the south of Tian He Lu and to the east of Guangzhou's urban central axis. Grandview Mall where the Aquarium is located boasts the largest singular commercial building in Asia, with the largest annual average sales and the highest visitor flow in China. The Aquarium is developed in the Mall to enhance the commercial atmosphere and attract the customers through unique experiences.

Grandview Aquarium, as the first in China developed through mall renovation, has the world's largest cambered acrylic display surface that is as wide as 40.8m. It occupies the southwest corner on F2 through F4 of Grandview Mall, at the former location of Friendship Store. The renovation involves 18,790m² while the mall area outside the Aquarium stays unchanged. The GFA of the Aquarium is 18,790m² (renovation area) comprising 7 floors, among which the renovation takes place on F2 through F4. Frame-shear wall structure is employed replacing the original frame structure.

The floor plan of the Aquarium is divided into 4 areas by function, namely visitor area, aquarium area, life-sustaining equipment area, BOH and evacuation area. As these areas are interdependent and interactive with each other, it requires repeated scrutiny to coordinate the specialties of architecture, structure, equipment, interior finishing, lift-sustaining system, acrylic display board installation, acousto-optic equipment, and operation management. In addition to complicated planar layout and circulation organization, another challenge to the renovation is the sectional design of space. Unlike building anew, the renovation was already restricted by the originally low floor height of 5m; not to mention the needs to reinforce the existing structure by adding a layer of floor slab that ranges from 400 to 600mm depth, which makes the space height even more awkward. Giving full consideration to the large load of the Aquarium, the structural renovation employs a number of new structural technologies to ensure the safety of the building, including: adding transfer girder to strengthen the cut column; providing new beam column strengthening node in the main tank's bottom plate area; adding steel plate shear wall and changing the original frame structural system to frame-shear wall structural system; adding damper and energy dissipation diagonal bracing; improving the ductility of building under earthquakes. The Project is a successful case in redeveloping and upgrading an existing commercial complex.

1

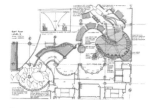

2

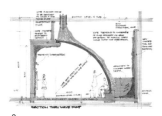

3

1 可与珍稀白鲸近距离对话
  Close encounter with rare white whale
2 总平面图
  Site plan
3 展缸细部
  Tank detail
4 270°超宽视野海底隧道
  270-degree super-wide field subsea tunnel
5 人造瀑布,营造欢乐刺激的气氛
  Joyful and exciting artificial waterfall
6 270°LED屏——营造星梦穿越的效果
  270-degree LED screen for a space travel effect

横琴创意谷
The Inno Valley Hengqin

横琴创意谷是省级创新基地、国家级重要项目；是横琴首个建成的大型办公及生态示范园区，是2015年珠海横琴十大重点工程之一。

横琴创意谷是设计全过程总承包的设计管理和服务。

设计理念是创享生态圈。平面通过退台和架空的手法，折线形体结合多层退台花园，体块搭接，内外廊连通，似平面内与外部景观交互渗透，形成灵动多变的空间效果，创造丰富的空间体验，同时，通过不同层数不同角度的旋转呼应场地内外景观节点，尽显空间趣味；立面设置间隔有序的竖向铝合金遮阳板，颜色丰富，形成独立的标志性立面风格，夜晚时分，缤纷柔和的灯光从其两侧折射而出，尽显金属的刚毅和玻璃的晶莹，现代感十足。整个园区通过生态轴线、生态庭院、生态水岸等一个个精美的小环境融汇出一个生机盎然的大环境，错落的庭院小景观，实现一步一景，步移景异的美妙体验。

项目地点	珠海横琴新区东侧，横琴新区核心地带，东临澳门大学
设计时间	2014.8
建设时间	2016.8
建筑面积	136668m²
建筑层数	地上5层
建筑高度	22.4m
合作单位	香港嘉柏建筑师事务所有限公司
获得奖项	2017年度广东省优秀工程勘察设计二等奖

As a provincial innovation base and national key project, Innovalley Hengqin is also a large office and ecological demonstration park firstly completed in Hengqin and one of the 10 key projects of Hengqin in 2015.

As the general contractor, we offer full-process design management and service for the Project.

The design concept is to create and share ecosphere. With various design approaches such as setback and open-up space, combination of folded form, multi-floor setback gardens, overlapped building volumes and interconnected interior and exterior corridor, the interior and exterior landscape are mutually penetrated to realize flexible spatial effect with diversified experience. Meanwhile, differently angled rotation on different floor responds to the internal and external landscape node, which makes the space more interesting. The colorful vertical aluminum alloy sunshade on regular interval presents the distinctive style of the façade. At night, illuminated by the flickering and soft light reflected from both wings, the building displays a modern look with metallic elements and crystal glazing. The ingeniously designed components including ecological axis, courtyard and waterfront jointly constitute the vigorous and exuberant park. The courtyard micro landscape is so arranged that is properly interspersed and varies as one moves to fascinate the visitors.

Location:	Core area and east of Henqing New Area with University of Macau to the east
Design:	2014.8
Construction:	2016.8
GFA:	136,668m²
Floors:	5 aboveground
Height:	22.4m
Partner:	Gravity Partnership Limited
Award:	The Second Prize of Excellent Engineering Exploration and Design Award of Guangdong Province, 2017

2

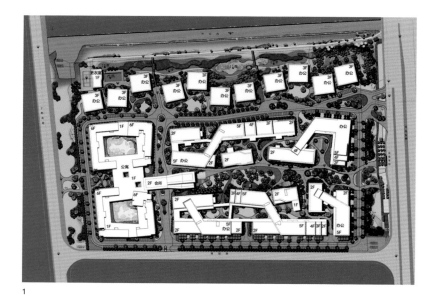

1

1 总平面图
  Site plan

2 园林庭院
  Garden courtyard

3 入口雕塑
  Entrance sculpture

4/6/7/8 立面局部肌理
  Partial façade fabric

5 平面布置灵动,体块互相咬合搭接
  Flexible planar layout with overlapped blocks

9 剖面图
  Section

S & T and Academic Buildings 科技教育建筑 **061**

商业、办公及酒店建筑

城市综合体是伴随经济飞速发展、建筑科学技术的进步、城市化水平的不断提升、土地资源日益紧缺而产生的一种建筑类型，也是一种使城市功能高度集约、延展了城市空间价值的建设和经营模式，它适应了城市化进程的多种需求。作为城市开发策略，它强调土地利用的均衡性；作为经济开发策略，它强调不同功能形成价值链的关系；作为城市空间设计策略，它强调空间集约和建筑形象的标志性。

商业城市综合体是在城市综合体的基础上，把商业功能作为综合体各部分功能的核心功能，把商业运营放在第一位，除了承担着城市综合体的公共利益外，商业价值和利益是商业城市综合体开发建设的主导因素。

高层商业城市综合体是指城市综合体中包含有高层商业、商务办公、酒店、公寓和住宅各类功能的高层建筑，以高层建筑主导建筑形态的城市综合体。高层商业城市综合体同时具备高层建筑及商业城市综合体的建筑特征，更能满足城市化发展的多种需求，是城市综合体中最为多见的一种建筑类型，它是适应社会集中高效、城市发展高层化的产物。

深圳万象城是GDAD早在2004年完成的一个高层商业城市综合体的代表作。当时，因为该项目的开发和设计的独特性，而获得了巨大的成功。它的空间和形态设计，非常强调商业氛围和人们进行商业活动的体验，在业态构成、流线组织、空间设置上都具有非常强烈的功能性。从城市空间和环境的角度出发，深圳万象城高层商业城市综合体的设计与城市空间和环境完美结合，为城市提供了一个高品质的公共空间。建成至今一直是中国商业城市综合体的标杆。

近年来，高层商业城市综合体发展的变化很快，业态构成呈现出复合的特征和跨界的趋向，在整体上则呈现出一种多元化商业发展模式。GDAD在高层商业城市综合体项目的建筑设计中，紧紧把握住建筑的时代特色，相继完成了惠州华贸中心商场、广州高德置地广场、保定万博广场、深圳招商局广场、广州绿地金融中心、深圳中广核大厦、昆明万达广场、佛山宗德服务中心等一系列的高品质的高层商业城市综合体项目，使高层商业城市综合体的建筑设计成为GDAD众多优秀设计品牌之一。

总结这些高层商业城市综合体项目的设计经验，我们得出这样的结论：建筑师应正确认识高层商业城市综合体与城市发展的关系，充分理解项目所在地域的文化、经济发展动向和项目的商业需求，根据用地地形特征和周围环境的条件，从整体和宏观的角度入手，统筹设计高层商业城市综合体的总体布局，合理布置各个功能分区，使建筑群体和城市集散空间有机结合，创造出建筑与城市的友好环境，让追求经济利益最大化的高层商业城市综合体能够充分地与城市文化、提升城市品位有机结合起来。

Commerce, Office and Hospitality Buildings

Urban complex, as a building typology, emerges along with the rapid economic development, advances in building science and technology, increasing urbanization level and shrinking land resources. It also represents a development and operation mode with highly intensive urban functionality and extended value of urban space, which is compatible with the varied demands of the urbanization progress. It gives priority to the equilibrium of land use as a strategy for urban development, emphasizes the value chain of varied functionality as a strategy for economic development, and focuses on intensified spaces and recognizability of the building as a strategy for urban space design.

Urban complex is the basis of urban commercial complex which incorporates commercial functionality as the core of all its functions, putting commercial operation at the first place. Apart from public interests undertaken by an urban complex, commercial value and profits are also the major factors leading the development of an urban commercial complex.

A high-rise urban commercial complex is an urban complex that takes the high-rises as main architectural form and is made up of the high-rises for retails, business office, hospitality, apartment and residence etc.. A high-rise urban commercial complex features the architectural properties of both the high-rise and the urban commercial complex, more capable of meeting varied demands of urbanization. As the most commonly seen architectural typology of urban complex, it is an architectural product that caters for the urban development toward a centralized, efficient and high-rise tendency.

The MixC Shenzhen is a representative high-rise urban commercial complex completed by GDAD in as early as 2004. At that time, the project was a great success due to its unique development and design. Its design of space and form emphasizes the commercial setting and experiences, highly functional in terms of trade mix, circulation organization and spatial layout. From the perspective of urban space and the environment, the design of The MixC Shenzhen perfectly integrates with the urban space and environment, offering a high-quality space for the city. Since its completion, The MixC Shenzhen has been an example of urban commercial complex in China.

During recent years, rapid changes have taken place in the development of high-rise urban commercial complex. Trade mix takes on increased complexity and the tendency of crossover, while on the whole, a diversified commercial development mode is taking shape. So far, GDAD, with a profound understanding of the time-related features of architecture, has designed a series of quality high-rise urban commercial complexes, including Huamao Central Place (Huizhou), GT Land Plaza (Guangzhou), Vanbo Plaza (Baoding), China Merchants Group Plaza (Shenzhen), Greenland Financial Center (Guangzhou), CGN Building (Shenzhen), Wanda Plaza (Kunming), German Service Center (Foshan) etc., making the architectural design of high-rise urban commercial complex one of GDAD's representative design brands.

A conclusion can be drawn by summarizing the design experiences in high-rise urban commercial complex, that the architects should have a correct understanding of the relation between the high-rise urban commercial complex and the urban development, and have an adequate knowledge of regional culture, economic development trend and commercial purposes of the project; meanwhile, the architects should plan the overall layout of high-rise urban commercial complex and rational functional zoning from an overall and macroscopic perspective based on geographical conditions of the site and surrounding environment, so as to integrate the building complex and urban gathering spaces, realize harmonious coexistence of architecture and the city, and combines the profit-oriented high-rise urban commercial complex with the urban culture and the improvement of urban taste.

索菲特酒店(圣丰广场)
Sofitel Hotel (Shengfeng Plaza)

广州天河商业圈兴建的广州索菲特圣丰广场,是一个集五星酒店(索菲特酒店)、甲级办公楼、商业等为一体的大型综合体项目。

酒店、办公塔楼相邻布置,共同围合出入门前广场。办公塔楼采用微纺锤形剖面设计,以椭圆形平面为蓝本逐层收分,高耸而挺拔。酒店以"L"形平面和写字楼组合,相互呼应,以流动的线条勾画出优美动感的造型,充分体现了现代建筑的特征,同时也塑造了优雅大气的整体形象。

索菲特圣丰广场为大底盘双塔楼结构,办公塔楼属高度超限建筑,框架柱全部采用大直径钢管混凝土柱;大底盘裙楼柱距均较大,局部柱距达20米以上;采用钢-混凝土组合结构新技术。另外项目还使用大量能减少对环境带来负荷的新技术、新设备,以实现绿化环境目标。如设计了可通风换气的玻璃幕墙和门窗,尽可能地提高建筑室内的舒适度,最大限度地满足非空调状态下的自然通风效果。办公部分采用冷凝水全热回收技术,把多余的能耗回收并利用起来。还有冷交换热泵供水系统、节水洁具、智能电气系统等,不仅提高了项目本身对于能源利用的效率,同时也减少项目营运的能耗和成本。

Located in Guangzhou's Tianhe business circle, the Sofitel Shengfeng Plaza is a large scale mixed-use development integrating five-star hotel (Sofitel Hotel), Grade A office building and commerce.

The hotel and office tower are placed next to each other, embracing the front entrance plaza. The office tower appears quite lofty thanks to its micro spindle-shaped section design and ellipse-shaped plan that shrinks by floor along the building height. The hotel in L-shaped plan joins and echoes to the office building with flowing lines and dynamic shape, fully reflecting the typical features of modern architecture and presenting an elegant image.

Sofitel Shengfeng Plaza has a large base with double towers. The office towers are over the code in terms of building height, with large-diameter concrete-filled steel tube columns as frame columns; while the large base podium employs large column grid of more than 20m at some positions. In addition to the new technology of steel-concrete composite structure, many up-to-date technologies and state-of-art equipment are applied to minimize adverse impact and foster green environment. For instance, ventilated glass facade, door and window are designed to improve indoor comfort level and maximize natural ventilation in non-air-conditioned period. Office areas are equipped with condensed water heat recovery technology for the recycle and reuse of excess energy. There are also cold exchanger heat pump water supply system, water-saving sanitary fixtures, and intelligent electrical system that can not only improve energy efficiency but also minimize energy consumption and cost during operation.

项目地点: 广东省广州市
设计时间: 2007
建设时间: 2008-2010.6
建筑面积: 154663m²
建筑层数: 酒店塔楼40层/办公塔楼28层
建筑高度: 184.5m
合作单位: Steffian Bradley Architects
曾获奖项: 2011年度广东省优秀工程勘察设计奖一等奖

Location: Guangzhou, Guangdong Province
Design: 2007
Construction: 2008-2010.6
GFA: 154,663m²
Floors: Hotel tower: 40; office tower: 28
Height: 184.5m
Partner: Steffian Bradley Architects
Award: The First Prize of Excellent Engineering Exploration and Design Award of Guangdong Province, 2011

1 在蓝天白云的映衬下格外耀目
  A dazzling presence against the blue sky and white clouds

2 总平面图
  Site plan

3 夜幕来临,暮色下的一缕璀璨
  A sparkling building in the dusk

1

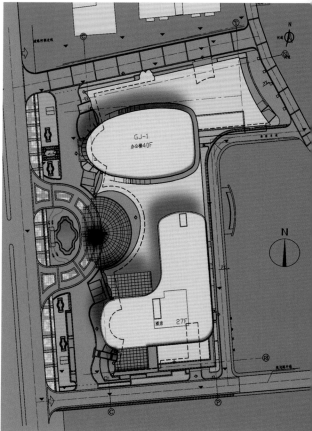
2

4 轻钢拱壳下的酒店大堂入口,倍感舒适
  Comfortable entrance to hotel lobby under a light steel arcade

5 屋顶恒温泳池
  Thermostatic swimming pool on roof

6 宽阔明亮的酒店大堂配上夺目的灯饰效果
  Spacious and bright hotel lobby with gorgeous lighting fixtures

7 舒适宜人的酒店标准间
  Comfortable standard guest room

Commerce, Office and Hospitality Buildings 商业、办公及酒店建筑 **067**

广东全球通大厦
Global Access Building

广州市珠江新城CBD核心区，为了给企业本身提供高效、健康的办公场所，并协调基地周边乃至整个区域的发展，决定兴建广东全球通大厦。工程总用地面积16,640平方米，由综合管理办公区(包括通信设备区)、会议展览中心、员工活动区、后勤服务及物业管理区等组成，总建筑面积约12万平方米。

设计借助CBD核心区的环境优势，整合内部空间资源，保证健康、舒适的室内、外环境，节约能源和资源，减少对自然环境影响，从总体规划布局、自然和生态环境影响、可再生资源利用、建筑围护结构、空调和采暖系统、照明系统等节能与建筑智能化管理技术等多个方面做了设计研究和应用。

项目于2010年9月广州亚运前竣工及投入使用，成了中国移动通信广东省的运营心脏，并顺利完成了作为2010年广州亚运会的通信保障指挥中心的任务。

项目地点：广东省广州市天河区珠江新城
设计时间：2006
建设时间：2010.6
建筑面积：121000m²
建筑高度：165.2m
建筑层数：37层
合作单位：加拿大黄雄溪建筑师事务所
曾获奖项：2011年度广东省优秀工程勘察设计奖二等奖

Location: Zhujiang New Town, Tianhe District, Guangzhou, Guangdong Province
Design: 2006
Construction: 2010.6
GFA: 121,000m²
Height: 165.2m
Floors: 37
Partner: Michael H.K. Wong Architects Inc.
Award: The Second Prize of Excellent Engineering Exploration and Design Award of Guangdong Province, 2011

The project aims to provide efficient and healthy office and coordinate the development of the surrounding area and beyond. Located at the Core Area of Zhujiang New Town CBD, Guangzhou, this 120,000m² development on 16,640m² site consists of office area (including telecom equipment area), convention and exhibition centre, and facilities for employee activities, BOH and property management.

In the design, the superior external resources in the Core Area are fully utilized to better organize the project's internal spaces, ensure the healthy and comfortable indoor and outdoor environment, and minimize the energy demands and impacts on natural environment. Besides, various design researches and applications are conducted in terms of site plan, impacts on natural and ecological environment, use of renewable resources, building envelop, HVAC, lighting and smart building system.

Being the operational center of China Mobile Group Guangdong Co., Ltd, the project was completed and put into use in September 2010 just before the Asian Games in Guangzhou. and functioned successfully as the communications security command center for the Asian Games 2010.

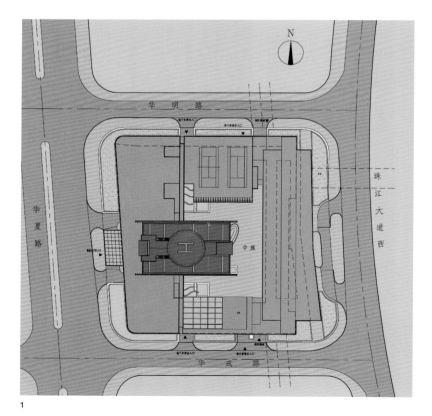

1

2

1 总平面图
  Site plan

2 整体东南角外景
  Southeast exterior view

3 鸟瞰全景
  Bird's eye view

江门电视中心
Jiangmen TV Center

作为江门地区大型电视中心，主要功能为电视事业的各类技术业务用房，包括录音制作用房、播音室、大型演播及电视专业设备用房、业务办公室等使用空间。项目位于广东省江门市北新区中心地段，主楼为22层，总高度97.5米，地下一层平时为机械停车库，战时为五级一等人防。2区裙楼功能包括三个演播厅、一个新闻报告厅，层数3层(1层夹层)，高度19.5米；3、4区裙楼功能包括网络中心、广告中心等技术、业务办公用房，层数5层，高度23.4米，总建筑面积49210平方米。

本工程所处城市环境地段重要，建筑造型既体现新时期广播电视建筑的时代特色，也成为江门这一现代化都市的标志性建筑物代表。电视中心内设有电视节目制作和播出、广播电视节目的接收和传输、数据业务的传输、网络管理、网络多功能业务的开发、监测管理、行政管理、资料及培训、广告、后勤与物业管理、用户服务管理等主要设施，配套仓库、水、电、空调用房、车库、停车场、广场和景观等设施。

电视中心满足江门电视台两套自办节目的制作和播出、38套模拟电视和60套数字电视以及10套调频广播节目的传输、数据业务传输及网络多功能开发动作的使用要求。

Jiangmen TV Center is a large TV center in Jiangmen area to house the main functions of various TV technical rooms, such as recording and production rooms, broadcasting studio, large studio and specialized TV equipment rooms, and offices. Located in the center of Beixin District, Jiangmen, the main building has 22 floors with a total height of 97.5m. B1 is used as a mechanical parking garage in peacetime, and a Level-5 Class-1 air defense in wartime. The 3-floor (1 mezzanine) 19.5m-high podium in Zone 2 includes three studios and one news broadcasting studio. The 5-floor 23.4m-high podiums in Zone 3 and 4 include rooms for technical and business use, such as the network center and ad center, covering a total floor area of 49,210m².

As the project is located in a key position in the city, the buildings not only reflect the architectural style of TV buildings in the new era, but also become a landmark in the modern cityscape of Jiangmen. The center is planned with main facilities for TV program production and broadcasting, radio and TV program receiving and transmission, data transmission, network management, development of multifunctional network services, monitoring and management, administration, materials and training, advertising, logistics and property management, and customer service management, as well as supporting facilities such as warehouse, rooms for water and power supply, air conditioning, garage, parking lot, squares and landscaping.

The TV Center is able to accommodate the production and broadcasting of the two programs by Jiangmen TV, and the transmission, data transmission and development of multifunctional network of the 38 analog TV, 60 digital TV and 10 FM radio programs.

项目地点：江门市北新区五邑大道
设计时间：2002.7-2004.4
建设时间：2004-2010
建筑面积：49210m²
建筑层数：地上22层，地下1层
建筑高度：97.50m
曾获奖项：2011年度广东省优秀工程勘察设计奖二等奖

Location: Wu Yi Da Dao, Beixin District, Jiangmen City
Design: 2002.7-2004.4
Construction: 2004-2010
GFA: 49,210m²
Floors: 22 aboveground, 1 underground
Height: 97.50m
Award: The Second Prize of Excellent Engineering Exploration and Design Award of Guangdong Province, 2011

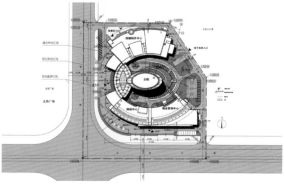

1 总平面图
Site plan

2 建筑各部分有机联系，采用园林绿化及水体布置，延续室内外空间
Various building parts are organically interconnected. The landscaping and water feature are provided to extend the interior and the exterior spaces

3 主楼采用观光电梯，对五邑广场做出有效呼应，在观光梯中内可看到五邑广场开阔空间和优美景观，电视中心同时也成为五邑广场的一道靓丽风景
The panoramic elevator in the main building effectively echoes to the Wuyi Square. Visitors may enjoy the open space and attractive view of the square from the elevators. The TV Center itself is also a part of the landscape of Wuyi Square

广东广播中心
Guangdong Broadcasting Center

在广州市人民北路,所处的地段集中了广州自新中国成立以来城市建设各个阶段的代表性建筑。广东广播中心体现了广州都市的建设成就,城市风貌和时代特色,同时也表现出广播事业的特点和岭南建筑特色,在环境构思方面,为城市增色。

广东广播中心工程地下二层,地上二十九层,总建筑面积67313平方米,建筑总高度99.95米,使用功广播专业技术用房及辅助设施、办公用房。该项目分近期、远期。

广东广播中心工程采用高科技手法,用金属材料低辐射玻璃作饰面材料,通过大量的现代化精细构件,组合多体量的不对称构图方式,避免建筑体型的过分庞大和呆板,同时,在设计上注重细部节点和材料的处理,将远期的高层塔楼置于近期主体的南端,使建筑主体的城市天际轮廓线轻灵秀丽,在体型组合上,以反复出现、互相呼应的弧面及曲线为主题,使建筑物造型丰富而形象统一,富有东方色彩。建筑细部处理在富于变化、尺度亲切的同时,亦有公共建筑的庄严感,整体色彩以纯净、淡雅为主,具有南方建筑的特点。

Located on Ren Min Bei Lu, Guangzhou, where one may find many representative buildings completed at various urban development stages since 1949. The project, as part of the urban development achievement, showcases views and features of the city and the gist of the times on one hand, and reflects the characteristics of broadcasting industry and Lingnan architecture on the other hand. It also enhances the urban appearance through landscape design.

The 99.95m-high building has two basement floors and twenty-nine above-grade floors, totalling a gross floor area of 67,313m². It houses functions of broadcasting studio, supporting facilities and office space. The project is divided into present and future phases.

The Center employs a hi-tech approach such as the cladding of metallic panels and Low-E glazing. Through a great number of modern fine components and the combination of multiple building volumes, it realizes an asymmetric composition which helps avoid the excessively massive or stiff architectural form. Meanwhile, the details, nodes and materiality are carefully devised. By placing the future tower at the south end of the present main building, a lively and elegant skyline is created. In combination of architectural forms, the curved cladding and curves are repeated as thematic elements to echo with each other, realizing the diverse yet consistent architectural form of oriental glamour. Besides, the building details of friendly scale, though varied, still contribute to a dignified image required for a pubic building. The building is in pure and elegant light colors, which is typical to architecture of South China.

项目地点:广东省广州市
设计时间:2001.4
建设时间:2001.6-2010.9
建筑面积:67313m²
建筑层数:29层
建筑高度:99.95m
曾获奖项:2013年度广东省优秀工程勘察设计三等奖

Location: Guangzhou, Guangdong Province
Design: 2001.4
Construction: 2001.6-2010.9
GFA: 67,313m²
Floors: 29
Height: 99.95m
Award: The Third Prize of Excellent Engineering Exploration and Design Award of Guangdong Province, 2013

1 总平面图
  Site plan

2 采用高科技手法,用金属材料低辐射玻璃作饰面材料,通过大量的现代化精细构件,组合多体量的不对称构图方式,避免建筑体型的过分庞大和呆板
  The Center employs a hi-tech approach such as the cladding of metallic panels and Low-E glazing. Through a great number of modern fine components and the combination of multiple building volumes, it realizes an asymmetric composition which helps avoid the excessively massive or stiff architectural form

3 避免建筑体型的过分庞大和呆板,在设计上注重细部节点和材料的处理
  To avoid excessively massive and stiff architectural form, details, nodes and materiality are carefully devised

4 人民北路沿街主立面整体色彩以纯净、淡雅为主,具有南方建筑的特点
  Street wall of Ren Min Bei Lu are mainly in pure and elegant light colors, which is typical to architecture of South China

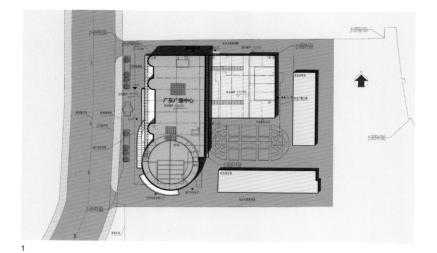

广州正佳商业广场东塔、西塔
East/West Tower of Grand View Plaza, Guangzhou

在广州市第一商圈——天河商圈的核心地段,天河路南侧,广州市中轴线东侧的广州正佳商业广场,是全亚洲单体规模最大的商业综合体,年均销售额及人流量全国第一。其项目包括的东塔、西塔是两座高层建筑。用地面积为2931.36平方米。两塔楼总建筑面积为132855平方米,建筑总高度:西塔182.3米(地上41层,地下4层及1夹层),东塔98.25米。

正佳广场综合商务大厦(西塔楼)是集超五星级酒店、酒店式商务房为一体的41层超高层建筑,在平面布局上以满足作为星级酒店客房所需建筑面积、模数的前提下,采用大柱网空间的形式,将建筑平面中部大胆设想、挖空处理,运用非常规的手法将建筑塔楼从10~25层和28~41层部分挖空,形成一整体性极强、气势磅礴的花园式中庭空间,从而建立建筑自身的个性、风格。将酒店客房、商务用房均设置在采光、通风良好的建筑平面周边,使每一套房均具有良好的风景视线区域,能摄取周边得天独厚的城市美景,更显示出建筑用地地理位置优越性。

东塔楼是集商业及写字办公楼为一体的25层高层建筑,地下一至五层布置商场,六至二十五层为写字楼,平面布局简洁合理,中筒位于平面中心,走廊围绕中筒而设,交通流线简洁方便,大大地提高建筑的实用性及采光面的利用率。

项目地点:广东省广州市
设计时间:2004.3-2010.10
建设时间:2011
建筑面积:132855m²
建筑层数:西塔41层;东塔25层
建筑高度:西塔182.3m;东塔98.25m
合作单位:广州晋泰建筑设计有限公司
曾获奖项:2007年度广东省优秀工程勘察设计奖二等奖(东塔)
第七届中国建筑学会优秀结构设计奖二等奖

Location: Guangzhou, Guangdong Province
Design: 2004.3-2010.10
Construction: 2011
GFA: 132,855m²
Floors: 41 (W); 25 (E)
Height: 182.3m (W); 98.25m(E)
Partner: Guangzhou Jintai Architectural Design Co., Ltd.
Awards: The Second Prize of Excellent Engineering Exploration and Design Award of Guangdong Province, 2007 (East Tower)
The Second Prize of the Seventh Excellent Structural Design Award by the Architectural Society of China

Located in the core of Tianhe Commercial Circle, the No.1 commercial circle of Guangzhou, on the south of Tian He Lu and to the east of Guangzhou's urban central axis, the Grandview Mall boasts the largest singular commercial complex in Asia and records the highest average annual sales and customer flow in China. The project includes two high-rises, namely the East Tower and West Tower. The two towers are planned with a site area of 2,931.36m² and a gross floor area of 132,855m². The West Tower measures a height of 182.3m tall with 41 above-grade floors and 1 mezzanine, and the East Tower 98.25m.

Grand View Plaza Business Tower (West Tower) is a 41-floor super high-rise complex integrating a super five star hotel and serviced apartments. Its floor plan, while providing the required floor areas and module of deluxe hotel rooms, employs large column grid to form central void from the tenth to the twenty-fifth floor and from the twenty-eighth to forty-first floor. With this novel approach, the magnificent garden atria are created in the central void, defining the identity and style of the building itself. Hotel rooms and offices are placed at perimeter of the floor plan to benefit from the favorable daylighting and ventilation. This also offers attractive city view to each room, further highlighting the prominent location of the project in the city.

The East Tower is a 25-floor high-rise complex including commercial facilities and office. A shopping mall is placed on the first five floors while the sixth to the twenty-fifth floors are office spaces. The floor plan is concise and reasonable, with the core at the center and corridors around. This enables simple and convenient circulation and greatly enhances the floor efficiency and the use of the daylit facade.

2

1

1 总平面图
  Site plan

2 天河路远眺的东塔实景
  View of the East Tower from Tian He Lu

3 简洁的造型,挺拔的姿态,相映成趣的双塔
  Two tower rise in concise forms and set off each other

深圳万象城
The MixC, Shenzhen

华润中心一期由一座国际5A甲级超高层写字楼"华润大厦"和一座大型室内购物中心"万象城"组成，位于中国深圳罗湖区地王大厦南侧，地处繁华的罗湖商业圈中心地段。设计通过引入地铁通道、设置下沉广场、建筑架空通廊、过街天桥、大型室内中庭等建筑手法，妥善地解决了项目的交通组织，并使建筑融入周边的城市环境中，为广大市民和观光旅游客提供了一个良好的购物旅游、休闲娱乐的场所，被评为"全国十大新地标综合体"。

华润大厦为地上29层超高层写字楼，地下3层为汽车库，地上总高度139.45米，建筑面积76260.3平方米；塔楼采用钢筋混凝土框－筒结构。华润万象城为地上5层高的商业娱乐中心，地下2层为商业、汽车库，地上总高度36.0米，建筑面积153612.8平方米。含大型真冰溜冰场、七厅电影城、超市、百货及各种专卖店、大小餐饮等。建筑平面外轮廓尺寸为198米X122米，单层面积超过2万平方米。设计以一个有天窗的半月形长条中庭形成交通主流线，并通过穿插布置的数十部自动扶梯，将五层商场和地下一层商场及地下二层车库联系起来。主体结构采用钢筋混凝土框架结构。

项目采用一系列的新材料、新技术和新工艺，如单索幕墙技术、金属钛锌板屋面、双银LOW－E中空钢化玻璃、分层钢管混凝土芯柱、RC节点施工技术、地下室超长无缝施工技术、大跨度钢桁架、渗透结晶体防水材料、大屋面虹吸排水、办公楼变风量空调系统等。

China Resources Center Phase One consists of "China Resources Building", an international 5A Grade A super high office building, and "The MixC", a large indoor shopping mall. The Project is located in the prosperous center of Luohu business circle south of Shun Hing Square in Luohu District of Shenzhen, China. Such architectural methodology as subway passage, sunken plaza, elevated corridor, overpass and large indoor atrium well addresses the traffic organization of the project while naturally integrating the buildings into the surrounding urban environment, providing an excellent place of shopping, tourism, leisure and entertainment for citizens and tourists. The project is honored as "Top Ten New Landmark Complex of China".

China Resources Building is a super high office building with 29 floors aboveground and 3 floors underground for parking. The total height above ground is 139.45m and the floor area is 76,260.3m^2; the tower employs reinforced concrete frame-core structure. The MixC is composed of a 5-floor aboveground commercial entertainment center and 2 underground floors for commerce and parking. The aboveground building totals a height of 36m and a floor area of 153,612.8m^2, including a large real ice skating rink, a seven-hall movie theatre, a supermarket, department stores and various franchised stores, large and small F&B etc. The size of building planar contour is 198mX122m with each floor being over 20,000m^2. A semilunar long atrium with skylight is designed for major traffic flow, and together with the distributed ten escalators, it links up the aboveground 5-floor shopping mall, the mall on B1 and the garage on B2. The main structure is reinforced concrete frame structure.

A series of new materials, technologies and processes are applied to the project, including single cable curtain wall technology, titanium-zinc metal plate roof, double silver Low-E hollow toughened glass, layered steel tube column filled with concrete, RC node construction technology, super-long seamless construction for basement, large-span steel truss, permeable crystallization waterproofing materials, large roof siphon drainage, and office building VAV air conditioning system etc.

项目地点：广东省深圳市罗湖区深南东路与宝安南路交汇处西南角
设计时间：2002.6-2003.6
建设时间：2002.10-2004.12
建筑面积：223418m^2
建筑层数：中区万象城5层，北区华润大厦29层
建筑高度：中区万象城23.9m，北区华润大厦139.2m
合作单位：RTKL
曾获奖项：2005年度全国优秀工程勘察设计行业奖二等奖
2005年度广东省优秀工程勘察设计奖一等奖
第四届全国建筑学会优秀结构设计二等奖
香港优质建筑大奖境外特别奖

Location:	Southwest corner of the crossing between Shen Nan Dong Lu and Bao An Nan Lu, Luohu District, Shenzhen, Guangdong Province
Design:	2002.6-2003.6
Construction:	2002.10-2004.12
GFA:	223,418m^2
Floors:	5 (The MixC in Middle Zone); 29 (China Resources Building in North Zone)
Height:	23.9m (The MixC in Middle Zone); 139.2m (China Resources Building in North Zone)
Partner:	RTKL
Awards:	The Second Prize of National Excellent Engineering Exploration and Design Award, 2005
The First Prize of Excellent Engineering Exploration and Design Award of Guangdong Province, 2005
The Second Prize of the Fourth Excellent Structural Design Award by the Architectural Society of China
Special Award for Building outside Hong Kong of the Quality Building Award

1

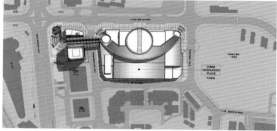

2

1 北区塔楼采用横向玻璃窗与竖向线条形成对比,使大体量的建筑不乏轻盈的姿态
Contrast between horizontal glass window and vertical lines on north tower, presenting a lightsome look of the large-volume building

2 总平面图
Site plan

3 鸟瞰图
Bird's eye view

4 深南大道实景
View of Shen Nan Da Dao

5 夜幕下灯光璀璨的万象城
The sparkling MixC at night

Commerce, Office and Hospitality Buildings 商业、办公及酒店建筑 **077**

6

7

8

9

6 连廊顶部作夸张处理,使连廊处成为视觉亮点,增加了"华润大厦"和"万象城"的整体性与连续性。
The exaggerated roof makes the corridor a visual impact, contributing to the unity and continuity of "China Resources Building" and "the MixC"

7 下沉式绿化广场与城市环境进行交流、融合,弱化城市环境与建筑边界,使建筑自然融合入城市环境。
The green sunken plaza interacts with and merges into the urban environment, blurring the boundary between urban environment and the buildings and incorporating the buildings naturally into the urban environment

8/9 "万象城"采用大体块,多元素、多材料组合的作法,给人以耳目一新的全新感受
The MixC features large building block and the combination of varied elements and materials, bringing about a refreshing building experience

11

12

10/12 中庭取意于购物街的概念，结合中庭特点，以较大的尺度呈弧线形展开，以丰富空间视觉效果
Inspired by the concept of shopping street in consideration of typical atrium features, the atrium unfolds like a large-scale arc creating diversified spatial visual effects

11 顶部饰以圆弧形采光天窗，为购物中心提供充足光源的同时，使人与自然和阳光有机会接触
The roof is decorated with arc-shaped skylight to provide sufficient light for the shopping mall and enable more contact of people with nature and sun

10

广州香格里拉大酒店
Shangri-La Hotel, Guangzhou

广州国际会展中心附近区域是广州新商业区的黄金地段，作为重要的商业设施而兴建的五星级酒店，广州香格里拉大酒店是广州首家真正豪华的国际酒店，毗邻广州国际会展中心，酒店可尽览珠江的秀丽风光，更有幽雅的翠绿庭苑，是尽享舒适和愉悦的都市中的一方绿洲，商旅及用餐的最佳选择。

广州香格里拉大酒店位于广州市海珠区新港东路，西邻会展中心，北朝珠江。以"亚洲式殷勤待客之道"为旅客们诠释香格里拉之典范，酒店设有客房750间、八间风格各异的餐厅及酒吧，两间宴会厅和八间多功能厅可满足宾客多样化的宴会需求。还有桑拿、室内外游泳池、网球场等配套设施。

建筑造型上采用高低组合方式，由37层的塔楼和裙房组合而成，塔楼高150米，建成后成为珠江南岸的重要城市景观之一。在建筑形式上，充分考虑了广州国际会展中心的建筑风格，建筑外立面与会展中心一样采用了玻璃和金属板，同样采用了富有流动感的曲线、曲面为造型元素，建筑外观简洁优雅、轻快而富有节奏感。

Prestigiously located at the area around Guangzhou International Convention and Exhibition Center (GZICEC), a prime location in the city's emerging commercial district, Shangri-La Hotel, Guangzhou is a five-star hotel developed as an important commercial facility. As the first truly luxury international hotel in Guangzhou, it is adjacent to GZICEC. With the attractive river view and a perfect blend of lush gardens and pavilions, it is truly a comfortable and pleasant oasis in the bustling city and the best choice for business travel and dining.

Shangri-La Hotel, Guangzhou is located at Xin Gang Dong Lu, Haizhu District, Guangzhou, neighboring GZICEC on the west and facing the Pearl River on the north. To serve its guests as a role model of "Asian hospitality", the hotel offers 750 hotel rooms, eight restaurants and bars of different styles, two ballrooms and eight multi-purpose rooms to cater for a variety of banquet needs, as well as other amenities including ballroom, sauna, indoor and outdoor swimming pools, tennis court and etc.

The building consists a 37-floor tower building on top of the podium. The 150m tower is one of the iconic views on the south shore of the Pearl River. The building form fully echoes with the architectural style of the Convention and Exhibition Center. Similar to the latter, the building uses glazing and metal panel for facades, as well as dynamic curved lines and surfaces to present a concise, elegant, brisk and rhythmic appearance.

项目地点：广州市海珠区新港东路广州国际会展中心东侧
设计时间：2003-2004
建设时间：2004-2007
建筑面积：122959m²
建筑层数：地上37层，地下2层
建筑高度：150m
合作单位：观光企画设计社集团（KANKO KIKAKU SEKKEISHA）
曾获奖项：广东省优秀工程勘察设计二等奖

Location: East of Guangzhou International Convention and Exhibition Center, Xin Gang Dong Lu, Haizhu District, Guangzhou
Design: 2003-2004
Construction: 2004-2007
GFA: 122,959m²
Floors: 37 aboveground, 2 underground
Height: 150m
Partner: KANKO KIKAKU SEKKEISHA
Award: The Second Prize of Excellent Engineering Exploration and Design Award of Guangdong Province

1 北向沿江立面，俯瞰珠江的秀丽景致
  Waterfront north façade overlooking the attractive river view

2 总平面图
  Site plan

3 南向立面，坐拥幽雅的青葱园林
  South façade overlooking lush gardens

惠州华贸中心商场
Huamao Center Mall, Huizhou

本项目位于广东省惠州市，总建筑面积150319平方米，是一个打破传统的束缚、具有地域特征的现代商场。

商场配套百货、BGH生活超市、粤东最大的3D数码影院、多个餐饮主力店等功能。

设计充分考虑了当地经济发展情况和业主开发需求，具有独特的建筑性格，具体表现为：为惠州江北新区CBD提供一个现代化，融合当地消费特点的商业巨擘；摈弃传统"1"字形动线，大胆创新结合三线城市发展实情的"p"字形动线，提高商场使用率；造型设计纯净大方，简练的体块组合与合理的材料搭配凸显了本案高档的定位，立面造型更显大气；室内空间不片面追求刺激和夸张，在营造浓厚商业气氛的同时兼顾人文关怀，为周边居民提供购物、餐饮、休闲、娱乐的舒适环境，体现以人为本的设计理念。

The project in Huizhou, Guangdong with a total floor area of 150,319m² boasts an unconventional modern shopping mall of regional characteristics.

The mall accommodates department stores, BGH supermarket, the largest 3D digital cinema in eastern Guangdong, and a number of anchor restaurants.

Taking full account of the local economic development and the needs of the client, the project has unique architectural characteristics as follows: standing out as a modern commercial mall integrating the characteristics of local consumption in the CBD in Jiangbei New District, Huizhou; Taking the p-shape circulation suitable for the development of tier-3 cities over the traditional 1-shape circulation to enhance the utilization of the mall; neat and elegant building form, combination of concise building volumes and elaborately devised materiality contribute to the high-end positioning of the project; instead of simply pursuing exciting and exaggerating design of interior space, attentions are given to creating a people-oriented commercial environment for the surrounding residents, satisfying their needs of shopping, dining, leisure, and entertainment.

项目地点：广东省惠州市
设计时间：2008
建设时间：2011
建筑面积：150000m²
建筑层数：5层
建筑高度：26.8m

Location: Huizhou City, Guangdong Province
Design: 2008
Construction: 2011
GFA: 150,000m²
Floors: 5
Height: 26.8m

1 简洁大方的主入口，尺度亲人，极具吸引力
Neat and elegant main entrance in agreeable scale is very attractive

2 总平面图
Site plan

3 鸟瞰角度充分体现建筑周边合理的交通组织
The bird's eye view shows the reasonable traffic organization around the buildings

4

4 灯光的独到运用为整个商场增加商业气氛
The unique lighting enhances the commercial environment

5-8 建筑细部力求完美,简约而不简单,从众多商业建筑中脱颖而出
With nearly flawless architectural details, the neat but not simple design of the project stands out from many of its peers

5

6

7

8

Commerce, Office and Hospitality Buildings 商业、办公及酒店建筑 085

东莞市商业中心区F区（海德广场）
Dongguan Commercial Center Zone F (Haide Plaza)

本项目的设计理念是创造"城市之门"与"没有背立面的建筑"的建筑形象，用"包容"的形象与城市景观和谐共融、刚柔并济，塑造东莞独一无二的五星级商务酒店和高档的写字楼，使其成为东莞的标志性建筑。建设地点位于东莞市新城市中心区，是东莞市商业中心区（又名：东莞第一国际商业区）的重点项目。其东北面为会展中心，西南面为购物中心，周边规划有轻轨站连接广州和深圳，地段得天独厚。

项目是一个集超五星级城市商务酒店、甲级写字楼、餐饮、会议、商业等功能为一体的大型建筑综合体，是一个顶部连体的超高层双塔建筑。以地块中轴线展开布局，建筑的中心位置位于原建筑群中轴线，总体对称。建筑与城市空间具有广泛的共融性，"双Y形"塔楼布局避免相互之间的视线干扰。建筑立面以"对称性"和"趣味性"相结合的姿态出现，以"对称性"对应于建筑群的轴线与整体城市空间，使建筑群与城市空间有很好的衔接并具有更广泛的关联性，以"趣味性"对应于商业建筑功能的多样性和复杂性。

场地以中央圆形广场为"太阳"中心，各景观元素及设施以"光环"形式向外顺序扩展。酒店大堂四层通高，面向中心花园广场，曲线形的空间轮廓，通透的玻璃幕墙，开阔的大堂共享空间，充分彰显出五星级酒店的超然气度。"Y"形平面拥有最小的核心、最大的采光面，标准层空间舒适紧凑，使用率高。塔楼在不同高度设有24层的空中庭园，作为休憩和景观空间，实现酒店房型和办公空间的多样化。本工程结构属A级高度的抗震超限工程，连体结构采用弧形钢结构桁架，共两榀，分别与两塔楼南、北两侧的外框柱连接，实现"城市之门"的建筑形象。

The project concept envisions an inclusive architectural image that not only highlights the idea of city gateway and building without back façade, but also harmoniously integrates the architecture with the cityscape. Designed with both strength and elegance, the building is a unique complex comprising five-star business hotel and high-end office building, hence a landmark in Dongguan. As a key project of Dongguan Commercial Center District (i.e. Dongguan First International Commercial District), it is favorably located in the city's new urban center, with the Convention and Conference Center to the northeast, shopping mall to the southwest and light rail station planned in the vicinity to connect Guangzhou and Shenzhen.

The project is a massive complex of twin towers that are interconnected at the top and integrates functions of super five-star business hotel, Grade A office building, F&B, conference and retail etc. The buildings layout is based on the central axis of the plot, with the center of the building falling right on the central axis of the original complex in a symmetrical layout on the whole. The architecture and urban space are widely integrated. The dual Y layout of the towers avoids mutual visual interference. The façade is both symmetrical and appealing. The symmetricity is relative to the axis of the building complex and the whole urban space, which enhances the relevance between the building complex and urban spaces. The appeal of the façade corresponds with the diversity and complexity of a commercial building.

With the central circular square as the "sun", i.e. the center, various landscape elements and facilities are configured sequentially in the form of "halo". The four-floor-high hotel lobby faces the central garden square. With curvilinear space profile, transparent glass façades and generous public spaces, it fully represents the grandeur of a five-star hotel. The Y-shaped floor plan realizes the minimum core, maximum daylighting, compact yet comfortable spaces on typical floors, and satisfactory floor efficiency. Sky gardens in 2 to 4 floor height are designed at various levels in the tower to provide lounge and landscape spaces and diversify the hotel room types and office spaces. The structure of the Project falls into Class A height seismic over-code project. Two curved steel trusses are used for the connecting structure to connect with the outer frame columns on south and north of the two towers respectively, presenting the desired architectural image of city gateway.

项目地点：广东 东莞
设计时间：2005-2008
建设时间：2006-2013
建筑面积：217000m²
建筑层数：地上37层，地下2层
建筑高度：158m
曾获奖项：2015年度全国优秀工程勘察设计行业奖建筑工程二等奖
2015年香港建筑师学会两岸四地建筑设计大奖卓越奖
2015年度广东省优秀工程勘察设计奖一等奖

Location: Dongguan, Guangdong Province
Design: 2005-2008
Construction: 2006-2013
GFA: 217,000m²
Floors: 37 aboveground, 2 underground
Height: 158m
Awards: The Second Prize of Architectural Engineering Design under National Excellent Engineering Exploration and Design Award, 2015
Excellence Award under Cross-Strait Architectural Design Awards by Hong Kong Institute of Architects, 2015
The First Prize of Excellent Engineering Design Award of Guangdong Province, 2015

1 方案设计手稿图
  Free hand sketch

2 总平面图
  Site plan

3 "东莞之门"的城市意象
  Image of "Gateway to Dongguan"

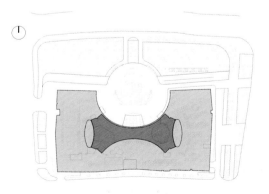

4	酒店大堂楼梯	Hotel lobby stairs
5	酒店侧庭	Hotel side foyer
6	酒店大堂	Hotel lobby
7	酒店主入口	Main entrance to hotel
8	设计手稿图	Free hand sketch

7

8

9 低点日景图
Low angle view at daytime

10-11 玻璃幕墙表皮
Building skin of glass façade

从化新城市民之家
Conghua New City Citizen Center

在过去几年里,广州实施了"北优"战略,推动了以从化市为核心的北部片区的蓬勃发展。从化将成为广州北部的城市副中心。城市需要"提质扩容",而从化新城的建设亦急需一个集中为市民政务服务的"平台",本项目在这样的背景下应运而生。

从化新城市民之家是新城的核心公共建筑,位于轴线河湖的北岸。复杂而多样的功能使得它本身就仿佛一个微缩的城市。设计提案将其整合为一个独立的城市体块,一个都市平台。都市平台连接着会议展览、综合审核、公共服务三大功能区,并将基础设施及公共服务配套等后勤功能合并成单一而有效率的整体。平台上设置连续的中心花园,同时向南、北延伸,成为"H"形城市花园,有机连接城市与市民之家,并与其他平台一起形成立体、动感的城市空间。在这里,多样化的对象及活动在同一时空发生和交织,叠加产生新的、绵延的城市活力。行人沿步道向中心聚集,到达南面首层宽阔的公共空间。这里也是地块的主要广场和城市的网络中心。室外步道可以通往展览中心,一条立体环形流线将引导观众进入各种展厅,了解当代最前沿各种资讯。流线自然延伸到屋顶花园,在此可一览新城全景。

市民之家是一个标志建筑,也是城市肌理的一个有机组成,它带给我们的不仅有标识性,还有亲切的社区和城市活力以及它们交叠而成的连续都市生态景观。在这样新型的都市综合体中,高效的都市营运和休闲多样化的都市生活合而为一。更多的公众参与使得市民之家不再是一个孤岛,而是具有多方连接的城市多面体,连续而开放的城市空间,犹如打通的都市经脉,最终为我们带来的,不只是短暂的灿烂,而是长久的活力。

项目地点:广州 从化
设计时间:2013年
用地面积:187000m²
建筑面积:240000m²
建筑层数:7层
建筑高度:28m
曾获奖项:2013年国内竞赛第一名

Location: Conghua, Guangzhou
Design: 2013
Site: 187,000m²
GFA: 240,000m²
Floors: 7
Height: 28m
Award: The First Place of domestic competition in 2013

With the implementation of Guangzhou's strategy of Optimizing the North, the northern part of the city, with Conghua at the core, has witnessed a vigorous development over the past few years. Conghua will become the future subcenter in northern Guangzhou. As the city demands quality improvement and expansion, the development of Conghua New City is also in need of a centralized platform for government services. The Project is planned in response to this need.

Located on the new city axis and the north bank of the river and lake, Conghua New City Citizen Center is a core public building with complicated and diverse functions, just like a miniature city. So the design proposal creates an independent urban block and platform for the project. This platform is connected to the three major functional areas, namely, convention and exhibition, review and approval, and public service; meanwhile, integrates the infrastructure, public service supporting facilities and other BOH and supporting spaces into one efficient whole. On the platform, the central garden continues and extends toward the north and south, forming an H-type urban garden with organic connection to the urban spaces and Citizen Home. The central garden, together with other gardens on the platform, contributes to vibrant multi-level urban spaces. Here, diverse objects and events take place simultaneously and interact with each other, generating new and ceaseless vitality for the city. The pedestrian paths are provided lead to the center and then a generous open space on the ground level south of the Citizen Center. It is the major plaza and urban network center on the site. Citizens may arrive at the Exhibition Center via the outdoor paths. A multi-level ring circulation will guide visitors into different exhibition halls for cutting-edge information of various fields. The circulation naturally extends to the roof garden where one may enjoy a panoramic view of the New City.

The Citizen Center is both a landmark and an integral part of the urban fabric. It brings not only an iconic image, but also amicable community, urban vitality and the interwoven and continuous urban ecosystem. In a new urban complex like the Citizen Center, efficient urban operation and diversified urban life become one within this novel urban space. Growing public participation makes the Citizen Center no longer an isolated island in the city, but a well-connected urban polyhedron and a continuous open urban space. Eventually, it will bring not only transient brilliance but also the long-lasting vitality.

1　总平面图
　　Site plan

2　市民中心夜景效果图
　　Night view of Citizen Center

3　与地形协调一致的绿化平台将三个体量串在一起,在空间上联系了三个不同的使用功能。剖面上深入考虑了生态、竖向交通整合的问题
　　The terrain-oriented greening platform links up the three volumes and spatially connects the three corresponding functions. The section incorporates considerations on ecological issues and vertical transportation integration

4　鸟瞰图:三个干净的体量嵌在新城中心,简约的体量蕴含了无限变化的内部空间
　　Bird's eye view: Three clear-cut volumes containing varied interior spaces are embedded into the center of the New City

5　整合、打开、提升、串联、引入、丰盈,一步一步生成从化新城的市民之家
　　The spaces are integrated, unfolded, upgraded, interconnected, introduced and enriched to gradually create the Citizen Center in Conghua New City

6　绿色平台沿H形花园延伸,直至城市广场,建筑和城市空间在此融为一体,形成连续的岭南都市生态景观
　　Green platform extends along H-shape garden till the urban square where the building and the urban space are integrated to form the continuous and ecological urban scene of Lingnan style

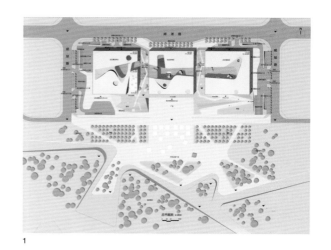

1

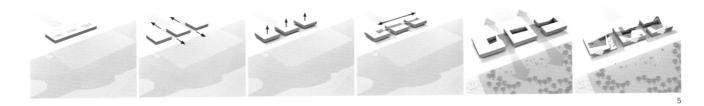

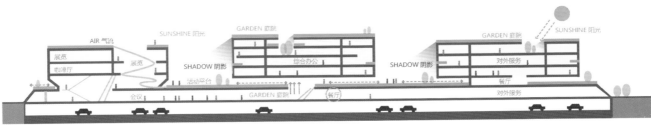

广州气象卫星地面B站区业务楼
Business Building of Guangzhou Meteorological Satellite Ground Station Zone B

建设方一直使用的气象卫星地面站A站区处于广州市市区内,并不适合气象卫星业务发展的需求。早在多年前,国家已计划在广州郊区另行建设B站区,通过多种因素比对确定了项目所在地。到目前为止,该站是全国仅有的四个气象卫星地面站之一,也是华南地区仅有的气象卫星地面站。项目包含了一系列建、构筑物,其中,业务楼除担任主要的技术任务外,局部区域还兼参观教育功能,因此业务楼在外形上最接近一般认知范围的建筑物形象。

众所周知,气象变化具有高度的复杂性与不确定性。而气象卫星业务中重要的一环是接收来自太空中气象卫星所拍的大气云图。业务楼建筑造型上并没有代入气象云图的具体形象,而是发掘其内在的复杂性与不确定性,以具备相当视觉冲击力的棱角和线条隐喻大自然气候的威力,建筑独特的外形成为B站区展示其工作复杂性和不确定性的最好代表。在强大的自然力量面前,人类要获得存在空间,必须协调与自然的关系,正如在如此独特的建筑外形下,内部空间如何保证满足各项功能一样。设计初期整合了所有条件,包括建设要求、使用要求、设施安装要求、施工工艺要求等,尽可能在不能正常使用的空间内化解形体和结构的矛盾以及安排设施或设备。

业务楼从协调和利用自然山风出发,结合调查岭南地区多年的生活经验,大多数公共空间并不安装空调系统,通过空间组织设计,在春夏秋季均可通过自然通风的方式达到一定的舒适度。同时考虑到冬季山风较冷的情况,多数主要活动空间均进行了节能设计。事实上,在业务楼装修并未完全完成的情况下,使用单位对局部区域舒适度的反映良好。

The Meteorological Satellite Ground Station Zone A that has been used by the client is in downtown Guangzhou and could not adapt to the development of meteorological satellite business. The plan to build Zone B in suburb Guangzhou was made years ago and finally the site was finalized through comparison. By now, Guangzhou Meteorological Satellite Ground Station Zone B is one of the only four meteorological satellite ground stations nationwide, and the only meteorological satellite ground station in southern China. The Project includes a series of structures, among which the Business Building mainly serve the needs of performing technical tasks, and particular areas are for educational purpose. Therefore the Business Building is similar to general buildings in form.

As one may know, climate change is highly complex and uncertain and one of the important works in meteorological satellite business is to receive atmospheric images taken by meteorological satellites from the outer space. Instead of showing the specific cloud charts, the Business Building tries to explore its inherent complexity and uncertainty to create corners and lines with strong visual impact, which metaphorically represents the power of nature. Its unique architectural shape can best reflect the complexity and uncertainty of the work being undertaken in Station Zone B. In the face of powerful forces of nature, human beings have to coordinate their own relationship with nature to have some space for living. It is the same to guarantee enough interior space for various functions within such unique architectural form. At the initial design stage, various project conditions are considered and coordinated, such as requirements for construction, usage, facility installation and construction technology, so that the conflicts between form and structure in ineffective spaces could be resolved and facilities or equipment be installed properly.

Without AC system in public spaces, certain comfort level of the Business Building could still be achieved through natural ventilation in spring and autumn as result of the spatial organization. In consideration of the cold wind in winter time, most activity spaces are designed in an energy efficient manner. In fact, even the fit-out of the building was only partially done, the user comments on the comfort level of the occupied spaces were quite positive.

项目地点:广州 萝岗
设计时间:2009-2010
建设时间:2010-2013
建筑面积:5539m²
建筑层数:地下1层,地上3层
建筑高度:13m
曾获奖项:广东省注册建筑师协会优秀建筑创作奖

Location: Luogang, Guangzhou
Design: 2009-2010
Construction: 2010-2013
GFA: 5,539m²
Floors: 1 underground, 3 aboveground,
Height: 13m
Award: Excellent Architecture Creation Award by Guangdong Chapter of Association of Chinese Registered Architects

1

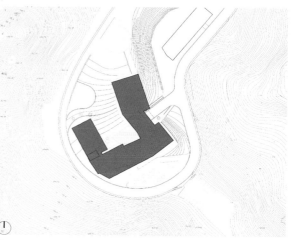

2

1 掩映在植物间的建筑一角
  Part of the building sheltered by green plants

2 总平面图
  Site plan

3 清晨的建筑外玻璃映照着东方鱼肚白
  Glazing façade reflecting the grey dawn twilight

4 从东南方向看建筑物,建筑与山体完美契合
  Perfect integration of architecture and mountain when viewed from southeast

5 设计手稿
  Design sketch

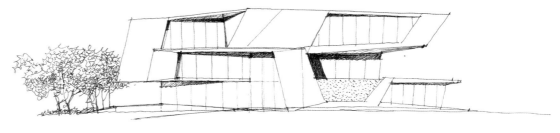

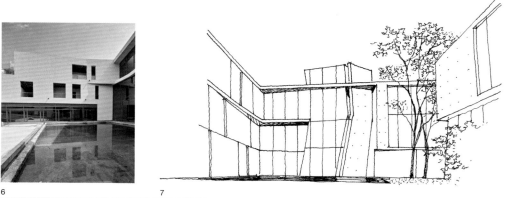

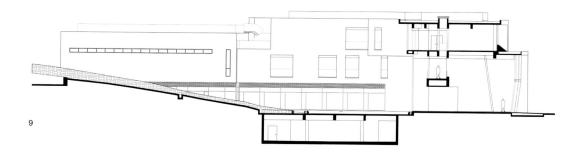

6	庭院以草坡及浅水景观池为两大主要元素，设计简洁，氛围静谧 With lawn slope and shallow landscape pool, the courtyard enjoys a tranquil and neat environment
7	庭院设计概念草图 Courtyard design sketch
8	雕塑感强烈的建筑立面局部 Sculpture-like façade
9	剖面图 Section

10/11　室外平台进深极大,压低了观赏视线,远方的山成了画面主角
The large depth of the semi-outdoor platform lowers the observation sightline and makes the mountain in a distance the key feature in the view

12　画廊式景框设计,把周边如画般的自然景色引入室内
Gallery-like landscape design frames and introduces the attractive natural landscape indoors

13　建设中的气象站
Meteorological station under construction

保利地产·珠海横琴发展大厦
Poly·Zhuhai Hengqin Development Mansion

珠海市横琴岛地处珠江口西岸，与澳门隔河相望，是带动珠三角、服务港澳、率先发展的粤港澳紧密合作示范区。横琴发展大厦是横琴新区的启动项目之一，依托横琴岛"山水相融"的规划布局，以"打造横琴特有的城市与建筑风貌"为主要目标，围绕三个最基本的要素展开设计：洁白光亮的建筑、遮阳的表皮与露台、风的竖井与架空层。

发展大厦塔楼方正，建筑高度达100米。位于二层的架空层及塔楼中央的"风的竖井"构成了建筑造型的最大特色，是实现其自然采光、通风的关键因素。架空层以下的建筑利用绿化屋顶做绿丘，延伸到南侧广场形成具有特色的大地景观。架空层以上的塔楼部分设置白色水平遮阳百叶，形成匀质界面。水平百叶与错位排列的开放式露台创造出轻盈灵动的建筑形象。塔楼内部功能布局集中，分区明确，流线明晰。核心筒四角均衡布置，使用者从各个区域都方便地到达目的地。平面空间可分可合，有利于适应不同类型的办公空间的灵活使用需求。周边绿化露台与室内办公空间、公共空间形成不同的组合，提供了有趣味、舒适的办公空间。

发展大厦的设计尊重岭南气候特征，注重舒适、节能，体现其智能化办公建筑的理念。架空层与风的竖井相结合，风沿绿丘而上，促进塔楼自然换气。水平遮阳百叶表皮与露台相结合，能最大限度地隔绝夏日骄阳，降低建筑能耗。此外，还使用雨水收集系统、太阳能光电板和光导照明等多项绿色建筑节能措施。发展大厦将成为先进的节能建筑物，达到全生命周期节能的绿色建筑目标，实现其可持续性。

Hengqin Island of Zhuhai, located to the west bank of the Pearl River Estuary and just a river away from Macao, is planned as a pilot area for close cooperation among Guangdong, Hong Kong and Macao. As the kick-off project of Hengqin New Area, Hengqin Development Mansion, under the master plan of integrating mountains and waters on Hengqin Island, aims to "create unique urban and architectural presence of Hengqin". The design centers on three basic elements, i.e. white and bright building, shaded skin and terrace, as well as air shaft and open-up floor.

The square-shaped mansion is 100m in height. The open-up floor on F2 and the air shaft in the center of the tower are the best highlights of the building and crucial factors for daylighting and ventilation. Green mounds transformed from the roof greening of structures under the open-up floor extend to the south square as part of the unique overall landscape. The white horizontal lamellas clad on the tower above the open-up floor and staggered open terraces jointly present a light and flexible building appearance. Internal functions of the tower are arranged in an intensive manner with clear zoning and circulation. Well-balanced configuration of the cores at four corners ensures easy access to all areas. The floor plan allows for flexible combination and subdivision hence can adapt to demands of different office typologies. The different combinations of greened terraces, interior offices and public spaces contribute to interesting and comfortable office spaces.

The project design takes into account the local climate features in Lingnan region, embodies the concept of intelligent office and emphasizes comfort level and energy efficiency. The open-up floor and air shafts make the upward air flow along the green mounds possible and enhance the natural ventilation of the tower. The horizontal sunshade lamella and terraces work together to maximize the solar insulation and energy efficiency in summer. In addition, the green building and energy efficiency measures such as rainwater harvesting systems, solar PV panels, optical conductor lighting are adopted. As a cutting-edge energy efficient building, the Development Mansion can attain the green building target throughout the life cycle with ensured sustainability.

项目地点：珠海 横琴新区
设计时间：2011-2014
建设时间：2011年至今
建筑面积：约230000m²
建筑层数：17层
建筑高度：100m
合作单位：日本佐藤综合计画
曾获奖项：2011年国际竞赛第一名
2014年广东省注册建筑师协会第七次优秀建筑佳作奖
2015年度广东省优秀工程勘察设计奖BIM二等奖

Location: Hengqin New Area, Zhuhai
Design: 2011-2014
Construction: 2011 to date
GFA: about 230,000m²
Floors: 17
Height: 100m
Partner: AXS
Awards: The First Place of International Design Competition, 2011
The Seventh Excellent Architecture Creation Award by Guangdong Chapter of Association of Chinese Registered Architects, 2014
The Second Prize of BIM under Excellent Engineering Design Award of Guangdong Province, 2015

1 总平面图
  Site Plan

2 景观绿化广场与低层绿化屋顶相连形成绿丘。绿丘之上悬浮着崭新的办公设施。开阔的开放性空间为公众活动提供优质的场所，大阶梯、二层天梯形成立体的多层次交通体系
  The landscaped square and the green roof of the lower floor are connected to form green mounds, with brand-new office facilities floating overhead. The open space offers quality venue for public activities, while the generous steps and sky stairs on F2 establish a multi-level traffic system

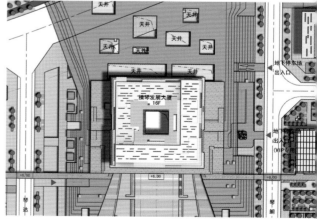

3

4

5

3 光塔式办公楼。夜晚,室内光线从遮阳百叶中透射出来,展现独具特色的反光效果,与广场、绿丘、水体的灯光相辉映,形成一番美景
Office buildings appear like light towers at night when the interior light penetrates from the sunshade lamellas, presenting an unique reflective effect. Coupled with the light in the square, green mounds and water body, a picturesque scene is then depicted

4 单体鸟瞰图
Bird's eye view of single building

5 立面设计强调有横琴风格建筑特色,采用白色基调、白色的横向百叶、错位展开的露台、四个立面不分主次,形成轻盈活泼的建筑立面形象
The elevation design highlights the Hengqin-style architectural features. The dominant white color, white horizontal lamellas, staggered terraces, and four equally important facades shape up an airily and lively elevation image

6 塔楼中央竖向天井——"风之竖井"与二层开放式架空层,共同构筑风的通道,为建筑内部带来自然的风和盎然的绿意
Central shaft (air shaft) and the F2 open-up floor jointly form an air channel that brings natural wind and abundant green into the building

美林湖度假酒店
Mayland Resort Hotel

在美林湖国际社区的核心区，美林湖度假酒店汇集了度假酒店、商务会议、宴会与餐饮、健身娱乐中心、水疗中心、高尔夫俱乐部等多种功能。基地坐拥美林湖国际社区山环水绕、空气清新的自然环境以及高尔夫球场等丰富的景观资源，加上与城市、机场之间便利的交通联系，是一个离世不离尘的世外桃源。

项目把地形、景观、功能、空间有机结合，整体造型变化丰富、美观，与周边环境协调统一。项目通过适应地形的轴线转折、层层叠叠的坡屋面、高低有致的楼台组合，形成"步移景异"、丰富多变的建筑群体，呈现建筑与土地和谐融合的原生形态。项目配合坡地建筑规划，设计强调古朴，师法自然，通过石材外立面、高低错落的陶质屋顶、喷泉、壁炉和庭院、铁艺、百叶窗，掩曳在藤蔓斑驳的墙头等的精心设计，创造单纯宁静的休闲度假环境，打造精雕细琢的欧洲小镇风情。

项目充分利用酒店靠近横坑水库的水源优势，采用"水源热泵＋离心冷水机组"的组合方式，夏季利用水源热泵的同时生产冷气和生活热水，冷气不足的部分用离心冷水机组补充，冬季利用水源热泵生产采暖和生活热水。冷水机组夏季制冷时，热泵机组冷凝器直接用湖水冷却，离心机采用冷却塔冷却；热泵机组冬季制热时，蒸发器从湖水取热。为保证水源热泵机组既能满足制热又能在制冷工况时的高效，分别采用以供热优先及以制冷优先两种工况的水源热泵机组。

项目地点：	广东省清远市
设计时间：	2007.6-2008.8
建设时间：	2008.8-2012.9
建筑面积：	64547m²
建筑层数：	山地建筑，上下共6层
建筑高度：	15.9m
合作单位：	美国JWDA建筑设计事务所
曾获奖项：	2013年度广东省优秀工程勘察设计奖三等奖

Location:	Qingyuan City, Guangdong Province
Design:	2007.6-2008.8
Construction:	2008.8-2012.9
GFA:	64,547m²
Floors:	mountainous building, 6 floors in total
Height:	15.9m
Partner:	JWDA
Award:	The Third Prize of Excellent Engineering Design Award of Guangdong Province, 2013

Mayland Resort Hotel is located at the heart of the Mayland International Community, serving a variety of functions including resort hotels, business meetings, banquets and catering, fitness and recreation centers, spas, and golf clubs. Boasting both rich landscape resources including the Mayland International Community by the water and mountains, the beautiful natural environment with clean air and the golf course, and convenient access to the urban area and the airport, the project is a convenient seclusion from the bustling city.

The project organically integrates the terrain, landscape, functions and space, while the varied and gorgeous building forms harmonize with the surrounding environment. With the terrain-based axis transformation, the slope roofs at various vertical levels and the well-organized buildings, building clusters with varied landscape are created to celebrate the harmony between the buildings and land. Furthermore, to be in line with the building planning for sloped site, the design emphasizes the simplicity and learns from the nature. Through the stone facades, the cascading ceramic roofs, fountains, fireplaces, courtyards, iron crafts, window blinds and wall covered in mottled vine, the project offers a tranquil and peaceful resort resembling a small European town.

Thanks to its close adjacency to the Hengkeng Reservoir, the project supplies cold air and hot water with water source heat pump units in summer (supplemented by centrifugal water chiller units in case of insufficient cold air), and hot water for heating and domestic use in winter. When the chiller units are working in summer, the condenser of the heat pump units are directly cooled with the lake water and the centrifuge by the cooling tower; when the heat pump units are working in winter, the evaporator takes heat from the lake water. In order to ensure high efficiency of cooling and heating, two types of water source heat pump units prioritizing heating and cooling respectively are provided.

1 总平面图
  Site plan

2 顺着地形层层跌落的建筑组合与层层跌落的水池形成建筑、景观的完美结合
  Cascading buildings along the terrain and the cascading pools perfectly integrate the architecture with the landscape

3 高低错落的建筑屋面与高低错落的室外园林相映成趣
　 The cascading building roofs form a delightful contrast with the outdoor garden at different vertical level

4 二层的主体建筑、一层的大雨篷以及高低错落的八角形、方形景观塔，形成丰富的入口空间
　 The 2-floor main building, 1-floor awning and the octagonal or square landscape towers of various heights jointly present an entrance of diverse view

海上世界酒店
Sea World Hotel

"东临碣石,面朝大海",深圳海上世界酒店位于深圳蛇口海上世界片区,与香港隔海相望,是由招商地产投资有限公司出资,并由国际著名的豪华品牌——希尔顿国际酒店管理公司在深圳打造的第一家豪华五星级商务度假酒店。"包容、优雅、自在"是其在现代主义外观下的人文气质:"硬朗"与"柔美"的墙身线条、"深灰"的玻璃幕墙与"暖灰"的石材幕墙形成强烈对比,在南国海滨谱写了一曲融汇"海浪、沙滩、椰林、岩石"元素的建筑"交响曲"。

设计运用现代的设计手法,结合现场特殊的地理特点和周边景观特征,独创性地把酒店入口和大堂区域提升至二层,由此展开,设置一系列公共区空间,如中西餐厅、特色餐厅、24H餐厅、室内恒温游泳池,SPA等呈线性融入一线海岸景观中。把酒店1200人的宴会厅和会议功能布置在首层,既减少了人流的穿插、满足了各部分功能使用的合理性,又最大化地利用了海景的资源。客房设计了两条弧形的塔楼向东合并延伸向海,寓意"双鱼",既营造了独一无二的造型设计又满足了323间客房全海景的需求。

深圳蛇口海上世界酒店还引入了多项节能减排措施,包括结合幕墙设计的可调节外遮阳、太阳能光伏照明、太阳能热水、中水回收、空调热回收等技术,被评为三星级绿色建筑。该项目在设计与施工过程中充分利用BIM辅助设计技术,最大限度地保证了设计与施工的一致性。

项目地点:深圳市南山区蛇口海上世界
设计时间:2010-2011
建设时间:2010-2013
建筑面积:56093m²
建筑层数:14层
建筑高度:63.35m
合作单位:美国WATG设计公司
曾获奖项:2014年度中国建筑学会优秀暖通空调工程设计三等奖
2015年度广东省优秀工程勘察设计奖二等奖
深圳市第十六届优秀工程勘察设计(公建类)一等奖

Location: Sea World, Shekou, Nanshan District, Shenzhen
Design: 2010-2011
Construction: 2010-2013
GFA: 56,093m²
Floors: 14
Height: 63.35m
Partner: WATG (US)
Awards: The Third Prize of Excellent HVAC Engineering Design by the Architectural Society of China, 2014
The Second Prize of Excellent Engineering Exploration and Design Award of Guangdong Province, 2015
The First Prize of the Sixteenth Excellent Engineering Survey and Design Award (Public Building) of Shenzhen

Facing the sea with stony hill on the eastern shore, Sea World Hotel is located in Seaworld Area, Shekou, Shenzhen, opposite to Hong Kong across the sea. As the first luxurious five-star business resort hotel in Shenzhen under the management of Hilton International Hotel Management Company, a world's renowned luxury brand, it is invested and developed by China Merchants Real Estate Investment Co., Ltd. The cultural quality of "tolerance, elegance and easiness" are revealed from the modern appearance: sharp contrast is created between the tough and gentle wall lines, and between the deep gray and warm gray stone facades, composing an architectural symphony featuring sea waves, beach, coconut trees and rockery by the coast of south China.

With the modern design approach and in view of the special site geographical features and the surrounding landscape features, the hotel entrance and lobby area are creatively raised to F2, followed by a series of public spaces. Specifically, the Chinese and western-style restaurants, featured restaurant, 24H restaurant, indoor constant-temperature swimming pool and Spa are integrated into the first-tier coastal landscape in a line. The 1,200-seating banquet hall and meeting facilities are provided on F1 to minimize the intersection of pedestrian circulations, ensure the reasonable functional layout and maximize the value of the seascape resources. The guest rooms are distributed inside two arc towers combined in the east and extended to the sea, symbolizing double fish, which creates a unique building form and makes the sea views available for all 323 guest rooms.

In addition, multiple energy saving and emission reduction measures are employed, including such technologies as adjustable external sunshades integrated with facade, solar PV lighting, solar hot water, recycled water recovery, and AC heat recovery, making the Hotel a three-star green building. In addition, BIM aided design technology are fully applied during the design and construction process to ensure the maximum compliance of construction with the design.

1

2

1. 总平面图
   Site plan

2. 酒店的西南面，酒店裙房及塔楼沿海岸线展开，充分利用了现场优美的自然景观资源
   On the southwest of the Hotel, the podium and tower extend along the coastline, making best use of the beautiful natural landscape

3. 酒店的北立面沿街透视，巨大的千人宴会厅和酒店塔楼的竖向交通体，完全独立出来，形成强烈的雕塑感
   Street-front perspective of hotel's north façade, the huge 1,000-seating banquet hall and the vertical transportation of the tower stand out completely with strong sense of sculpture

4. 酒店的东北角主入口一侧，优美的园林景观把酒店的主入口引至二层中部，舒展的酒店塔楼客房完全朝海岸线展开
   Along one side of the main entrance at the northeast corner of the Hotel, attractive garden landscape introduces the main entrance to the middle of F2, and guestrooms in the tower of the Hotel completely stretch towards the coastline

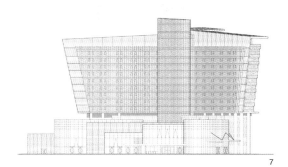

5 酒店二层平台景观水池透视图，打造出海天一色的美景
F2 reflective pool sets off with the spectacular view of the sea and sky.

6 入口低点透视
Low viewing point entrance perspective

7 立面图
Elevation

8 酒店入口局部透视图
Perspective of hotel entrance

9 宴会厅
Ballroom

10 全日餐厅内景
Interior of 24H restaurant

白云绿地金融中心
Baiyun Greenland Financial Center

2003年老白云机场搬迁后规划的白云新城，发展定位调整为"广州市城市副中心，白云区区级中心，现代生化生态型商贸文化中心。"该区域地处广州市西北部，白云区的南部，紧靠中心城区，是新机场进入广州中心城区的门户地区。本项目地处白云新城核心区域，即老白云机场跑道位置，现已成为该区域的地标性建筑。

本项目建筑裙楼部分强调与白云山山体状体块相协调，充分考虑路面与行人对建筑体量的认知，在沿市政路的外侧采用相对大的体量设计，给人以整体的感受；而在场地内侧把体量打碎，以小体量的裙楼与架空连廊相结合，丰富空间形式。裙楼融入旧白云机场的登机平台、廊桥等建筑元素，体现对历史记忆的延续。连接西北侧地铁的下沉广场与北侧城市绿化带结合为一体，同时通过裙楼西北角的退台式设计，将绿色引向建筑主体；多维度地创造出富有生趣的生态休闲商业场所。塔楼由下到上逐层收分，形式典雅，完美展示出建筑形体。

本项目从概念设计到规划、单体充分体现了对白云山周边环境的理解与尊重，建筑裙楼形体通过层层退台与屋顶和垂直绿化的运用，使整栋融入周边环境。地下空间的开发与利用，通过多个下沉广场的设置，解决部分地下空间的采光通风问题，同时又给消防设计提供疏散空间。下沉广场的设置并与地铁连接，它不仅成为本项目的交通集散节点，更成为了城市空间与轨道交通的中转站。在设计过程中运用BIM技术着重进行管线综合设计，以解决设计阶段各设备管线与建筑、结构的空间关系，进而为现场施工提供直观的施工指导，以降低施工难度和反复性。最终为投资方降低投资成本和建设周期提供条件。通过融入绿色建筑的设计，提高使用品质，有效降低运营成本。

项目地点：广州市白云新城
设计时间：2011-2012
建设时间：2012-2013
建筑面积：292183m²
建筑层数：地下4层，地上46层
建筑高度：200m
合作单位：株式会社日本设计
曾获奖项：2015年度全国优秀工程勘察设计行业奖 建筑工程二等奖
2015年度全国优秀工程勘察设计行业奖 建筑结构一等奖
2015年度广东省优秀工程勘察设计奖一等奖
2015年度广东省优秀工程勘察设计奖 建筑结构专项一等奖

As planned after the relocation of the old airport in 2003, Baiyun New Town is repositioned as a subcenter of Guangzhou, center of Baiyun District and a modern and eco-friendly commercial and cultural center. It lies in the northwest of Guangzhou and the south of Baiyun District, with close adjacency to downtown, boasting the gateway from the new Baiyun International Airport to downtown Guangzhou. The Project is located in the core area of Baiyun New Town, and has now become a landmark of the area.

The podium of the project highlights and harmonizes with the form of Baiyun Mountain, with careful consideration of pedestrians' cognitive awareness of building volume. Relatively large building volume is adopted along municipal roads, ensuring a overall building presence, while smaller building volumes are provided inside the site, which are connected with open air corridors on podium level, providing various spatial forms. Boarding platform and covered bridge and other architectural elements used in former Baiyun Airport are employed in podium to remind visitors of the historical memory of the site. A sunken plaza leading to metro to the northwest of the plot is merged into the urban green belt to the north of the site and the recessed platform at the northwest corner of the podium brings greening into the main building, creating a fun and eco-friendly place for leisure and commerce in various ways. The tower tapers from bottom to top, presenting an elegant architectural form.

The project shows full understanding of and respect for the Baiyun Mt. and the surrounding environment all through the stages of concept, planning and singular building design. Recessed platforms and roof and vertical greening of the podium enable the whole building well blended into the context. The development and utilization of underground space is addressed via multiple sunken plazas that tackle the daylighting and ventilation of partial underground spaces and offer fire evacuation space. The sunken plazas are provided to connect with the metro line as the traffic node of the project and the transit point between urban space and rail transportation. BIM technology is adopted for integrated piping design to address the spatial relationships between pipelines, building and structure at various design stages and to provide clear instructions for construction. This helps reduce difficulty and repetitiveness of construction and eventually cut investment and construction period. Besides, the green building design is incorporated to enhance the quality and cut operating costs.

Location: Baiyun New Town, Guangzhou
Design: 2011-2012
Construction: 2012-2013
GFA: 292,183m²
Floors: 4 underground, 46 aboveground
Height: 200m
Partner: Nihon Sekki Inc (Japan)
Awards: The Second Prize of Architectural Engineering under National Excellent Engineering Exploration and Design Award, 2015
The First Prize of Building Structure under National Excellent Engineering Exploration and Design Award, 2015
The First Prize of Excellent Engineering Exploration and Design Award of Guangdong Province, 2015
The First Prize of Building Structure under Excellent Engineering Exploration and Design Award of Guangdong Province, 2015

1 总平面图
  Site plan

2 裙楼层层退台凸显层次感
  The podium sets back by floor, enhancing the sense of layers

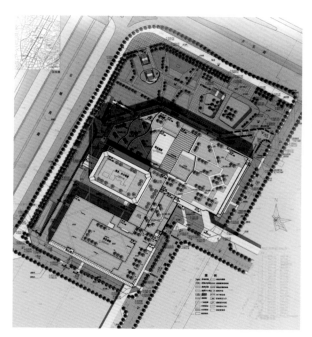

1

3 塔楼逐层收分、富有变化的玻璃幕表皮
　Tapering and varied tower facade

4 外挑金属大雨棚强调力与美的结合
　The generous cantilevered metallic canopy highlights the combination of strength and aesthetics.

5 二期商业细部
　Details of Phase II commercial wing

6/7 塔楼办公大堂
　　Office lobby of the tower

Commerce, Office and Hospitality Buildings 商业、办公及酒店建筑　**109**

招商局广场
China Merchants Group Plaza

30多年前,蛇口是这个时代的起点,改革开放从这里走向全国,创造了举世瞩目的奇迹;这里留有法兰西总统戴高乐乘坐过的"明华轮";这里有邓小平亲笔题字的"海上世界"……这里承载了太多荣耀和记忆。如今,招商局广场的建成,成为蛇口片区最高的标志性办公建筑,整栋建筑物造型挺拔,个性沉稳不张扬,体现了顶级商务办公楼的优雅气质,同时充满了现代感和前瞻性,是蛇口大南山区域天际线中的亮点及转折点。

高18米大堂,4.5米标准层高,智能电梯预约系统等,超前配量为深圳蛇口自贸区最高端超甲级写字楼;顺应立面造型,为获得更大的使用面积和经济效益的需要,整栋塔楼除了核心筒以外由16根周边柱自首层到11层外倾2.23°,之后到顶层的部分内倾1.27°,所产生的楼层面积随着造型的收分而变化;在建筑的四个立面设计上采用竖向线条来强化建筑垂直品质,外立面设计在平面高效、结构合理的前提下,综合了海、帆、灯塔等元素的外部造型,表达了一股向上的精神,招商局广场成为蛇口新一代的标志和象征。

建筑的外围护是由定制挤压成形铝制单元式玻璃幕墙构成。外层玻璃是高性能的Low-E中空玻璃,根据需要进行热强化。在不透明的楼层窗间墙位置处结合单元式玻璃幕墙,设置岩棉隔热层,并包括带有高性能喷涂铝合金保温衬板的透明平板玻璃。在建筑玻璃幕墙的垂直方向,每隔约0.9米设0.35米宽、白色不透明夹胶玻璃百叶的水平遮阳,既丰富了建筑外立面的造型,又对建筑的节能有利。

Three decades ago, Shekou was the start of a new age, from where the reform and opening up policy went nationwide and created a remarkable miracle. It holds many glories and memories, including Minghua Ship that once carried French president Charles de Gaulle, and the Sea World with Deng Xiaoping's handwritten inscriptions. Today, Shekou is marked by China Merchants Group Plaza, the tallest iconic office building in the area with lofty shape and calm personality. With the elegant posture of a top-grade business office building and highly modern and forward-looking style, China Merchants Group Plaza stands as a highlight and turning point on the skyline of Shekou in the larger Nanshan District.

Advanced configurations including 18m-high lobby, 4.5m typical floor height and intelligent elevator system make it the best end super Grade A office building in Shekou Free Trade Zone of Shenzhen. To fit the façade form, the entire tower except the core is designed with 16 columns on the perimeter inclining outward 2.23° from the first to the eleventh floor and then inward 1.27° up to the top floor. The area of each floor changes with the expansion and contraction of external shape. Facades on all four sides are decorated with vertical lines to reinforce the verticality of the building. Based on an efficient plan layout and a rational structure, the facade also incorporates elements including sea, sail, light house etc., presenting a spirit of striving upward and a new icon of Shekou.

The building is enclosed by custom-made extruded aluminum unit glass facade, whose external layer features high performance Low-E hollow glass that can be thermally toughened as necessary. Rock wool thermal insulation layer, including transparent sheet glass with high performance coated aluminum alloy insulation plate, is installed in the opaque wall between windows in consideration of the unit glass facade. Horizontal sunshades using 0.35m wide white opaque laminated glass louvers are installed every 0.9m vertically along the glass facade, creating diversified façade form while achieving energy efficiency.

项目地点:深圳市南山区蛇口海上世界
设计时间:2008.9-2010.4
建设时间:2010.12
建筑面积:107275m²
建筑层数:37层塔楼和3层裙房
建筑高度:211.01m
合作单位:美国SOM建筑设计事务所
曾获奖项:2013年中国建筑学会优秀结构工程设计银奖
2014年中国建筑学会优秀给排水设计三等奖
2015年度全国优秀工程勘察设计行业奖二等奖
2015年度广东省优秀工程勘察设计奖一等奖
深圳市第十六届优秀工程勘察设计(公建建筑)一等奖

Location: Sea World, Shekou, Nanshan District, Shenzhen
Design: 2008.9-2010.4
Construction: 2010.12
GFA: 107,275m²
Floors: tower: 37; podium: 3
Height: 211.01m
Partner: SOM
Awards: Silver Award of Excellent Structural Engineering Design by the Architectural Society of China, 2013
The Third Prize of Excellent Plumbing and Drainage Design by the Architectural Society of China, 2014
The Second Prize of National Excellent Engineering Exploration and Design Award, 2015
The First Prize of Excellent Engineering Exploration and Design Award of Guangdong Province, 2015
The First Prize of the Sixteenth Excellent Engineering Survey and Design Award (Public Building) of Shenzhen

1 总平面图
  Site plan

2 招商局广场片区概貌
  A view of China Merchants Plaza area

3 夜色中的办公塔楼,犹如灯塔般耀眼
  The office tower resembles a brightly lit beacon at night

1

2

Commerce, Office and Hospitality Buildings 商业、办公及酒店建筑

4 多面体的办公塔楼
　Polyhedral office tower

5/6 办公塔与商业裙房相连的视觉连廊
　　Visual corridor connecting office
　　tower and commercial podium

4

5

6

7/8 办公大堂内部设计
　　 Interior design of office lobby

9　1616平方米恢宏入户大堂，17.9米层高，四面通透采光，彰显奢华气度
　　Lofty entrance lobby of 1,616m² large and 17.9m high allows for transparent view on all four sides

Commerce, Office and Hospitality Buildings　商业、办公及酒店建筑　113

中广核大厦
CGN Building

在深南大道,由东往西进入深圳市CBD(中央商务区)中心区的门户位置,中广核大厦位于深圳市福田区深南大道和彩田路交口处东北角,项目西侧是区内道路、北侧是福中三路、东侧是彩田路、南侧是深南大道。用地功能为商业办公用地,建筑性质为商业性办公楼,建筑内容包括39层南塔楼、24层北塔楼、3层地下室和3层裙楼。

外观设计上,超高南北双塔造型交相呼应、空中围合,形成"L7"的建筑形象,与项目的CBD门户位置相呼应;1.8米小平窗、2.4米中凸窗、3.0米大凸窗所形成的凹凸窗墙体系,简洁中蕴藏着微妙变化,参数化排列形成栉比鳞次的起伏状;搭配上较深沉色彩的金属质感外墙,整个大厦建筑远观有清晰简明的形体轮廓,近观又有丰富多变的肌理层次,充分展示着中广核集团作为未来国际一流核电企业所特有的严谨、稳健、大气恢宏的形象。

功能设计上,项目从经济、实用角度,融入了对内部空间利用、人车分流、节能日照等诸多要素的考虑,较好地满足了集团公司总部办公大楼作为高档写字楼对内部办公、科普展览及综合配套的功能要求。北楼屋顶太阳能光伏发电系统、南楼屋顶直升机停机坪等现代高科技的引进,使大厦更添时代先进色彩。

Located in the northeast corner at the intersection of Shen Nan Da Dao and Cai Tian Lu in Futian District of Shenzhen, CGN Building is the gateway to the central area of Shenzhen CBD (Central Business District) when accessed from east to west along Shen Nan Da Dao. Defined by an internal road on the west, Fu Zhong San Lu on the north, Cai Tian Lu on the east and Shen Nan Da Dao on the south, the site is planned for commercial office purposes, accommodating commercial office buildings, including a 39-floor south tower, a 24-floor north tower, a 3-floor basement and a 3-floor podium.

In terms of building appearance, the super high south and north towers embrace each other into a shape of "L7", which echoes the Project positioning as a gateway to the CBD; the concave-convex window system consisting of 1.8m small flat window, 2.4m medium-sized bay window and 3.0m large bay window looks concise with subtle changes, all aligned orderly into the rise and fall of serrated rows; with metallic exterior wall of dark colors, the entire building complex presents a distinct and concise profile when seen from a distance but rich and variable textures and layers when one comes closer, all to exhibit a rigorous, solid and grand public image of CGN Group as a world-leading nuclear power enterprise in the future.

In terms of functional design, the project adopts an economical and practical approach that considers various factors, including the use of the interior spaces, separation of pedestrians from vehicular circulation, energy conservation and daylighting etc. This approach properly meets the demands of a high-end headquarters office building for offices, popular science exhibition and comprehensive supporting facilities. Modern technologies including roof solar photovoltaic power generation system on north building and roof helipad on south building keep the buildings up with the times.

项目地点:深圳市中心区东北(深南大道与彩田路交界处西北角)
设计时间:2009.10-2010.3
建设时间:2010年至今
建筑面积:158830m²
建筑层数:地下室3层,地上北楼24层,南楼39层
建筑高度:北楼99.97m,南楼176.9m
合作单位:都市实践设计有限公司
曾获奖项:2013年中国建筑学会优秀结构工程设计银奖

Location: Northeast of Shenzhen's Central Area (northwest corner of the crossing between Shen Nan Da Dao and Cai Tian Lu)
Design: 2009.10-2010.3
Construction: 2010 to date
GFA: 158,830m²
Floors: Underground:3, aboveground: 24(N) and 39 (S)
Height: 99.97m (N) 176.9m (S)
Partner: URBANUS
Award: Silver Award of Excellent Structural Engineering Design by the Architectural Society of China, 2013

1 总平面图
  Site plan

2 简明清晰的建筑形象在中心区纷杂的建筑群中脱颖而出
  Concise and distinct building image stands out from the chaotic building clusters in the central area

3 远观有清晰简洁的形体轮廓 近观有丰富多变的肌理层次
  A distant view shows a distinct and concise building profile, while a close-up look reveals rich and variable textures and layers

4/5 形体在平面和空间上相互交错,形成咬合互动的整体形象
  Building volumes intertwine in both plan and space, presenting an interlocked and interactive image

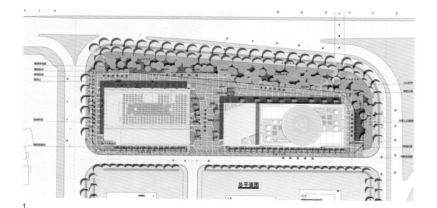

Commerce, Office and Hospitality Buildings 商业、办公及酒店建筑 115

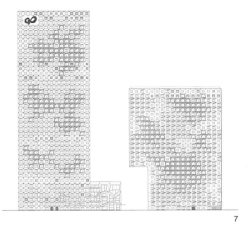

6	形体在平面和空间上相互交错,形成咬合互动的整体形象 Building volumes intertwine in both plan and space, presenting an interlocked and interactive image
7	立面图 Elevation
8-10	建筑立面细部,窗形的变化带来丰富变幻的光影效果 Facade details and change of window style bring out diversified light and shadow
11	公共大堂色调与肌理延续外墙风格,营造简洁明快的办公空间 The public hall continues the color and fabric of the exterior wall, creating simple and bright office spaces

海上世界广场 – 船后广场
Seaworld Plaza - Ship-back Square

海上世界广场是由邓小平同志题字的深圳文化地标,作为改革开放集体记忆场所,更新后成为深圳南山蛇口的标志性建筑。海上世界广场—— 船后广场作为海上世界"明华轮"环船商业服务项目之一,目标成为蛇口片区乃至南山、深圳市区市民旅游、休闲、购物等使用的舒适、宜人的商业项目。

船后广场位于"明华轮"及中心湖面的东南面,建筑公共空间(商业动线)完全开放联系城市空间,融入蛇口街区肌理,实现高度的通达性、公众性。设计充分考虑蛇口的气候特征,吸收岭南建筑生态环境设计理念,结合项目的休闲氛围,注重建筑节能和室内外生态环境设计,利用建筑或技术手段解决自然通风、日照控制等问题,采用低能耗的先进设备营造宜人的小气候。建筑造型理念是以反映海洋文化的设计风格体现本区特有的人文和环境特点,"双鱼"建筑形态虚实对比、体量互相咬合,表现出"鱼跃欢腾"的动态;相互穿插的连廊动线空间,犹如律动的琴响,欢愉的体验感染着每一个到来的市民。

建筑的外围护部分采用最新的系统和表皮材料,以一种更加现代的方式表述连续转变的屋面墙面形态,生动表现自然流动的特性。表皮的实体部分采用高品质的白色亚克力外墙材料,可加入LED背光,白色半透明及反光度加上LED的应用在整个建筑上,使之富于神秘感。建筑的半透明部分采用异形半透明彩釉玻璃附着在优雅的空间框架结构上,覆盖广场的同时为露台餐饮提供独特的采光介质。

项目地点:深圳市南山区蛇口海上世界
设计时间:2010-2011
建设时间:2011-2013
建筑面积:29986.9m²
建筑层数:3层
建筑高度:16m
合作单位:美国凯里森建筑设计事务所
曾获奖项:深圳市第十六届优秀工程勘察设计(公建建筑)一等奖

Location: Sea World, Shekou, Nanshan District, Shenzhen
Design: 2010-2013
Construction: 2011-2013
GFA: 29,986.9m²
Floors: 3
Height: 16m
Partner: CALLISON
Award: The First Prize of the Sixteenth Excellent Engineering Survey and Design Award (Public Building) of Shenzhen

With its name inscribed by the late state leader Deng Xiaoping, the Seaworld Plaza is the cultural landmark of Shenzhen and the place that carries the city's collective memory about reform and opening-up. After the renewal, it has become a representative building in Shekou. Its Ship-back Square, as one of the commercial service projects around the Minghua Ship, is planned as an attractive and pleasant destination for sightseeing, leisure and shopping of citizens in Shekou, Nanshan District and even the whole city of Shenzhen.

Ship-back Square lies to the southeast of "Minghua Ship" and the central lake. Its public spaces (commercial circulations) are completely open to connect with urban space and integrate into the neighborhood context of Shekou for high accessibility and publicity. With full consideration to the climatic characteristics of Shekou and incorporation of the eco design concept of Lingnan architecture, we put emphasis on building energy efficiency and indoor and outdoor ecological environment in view of the leisure atmosphere of the Project, realize natural ventilation and sunlight control through architectural or technical approaches and create favorable microclimate with cutting-edge low-energy equipment. The building shape is conceived to reflect the maritime culture and the unique cultural and environmental characteristics of the region. The double-fish building form features a comparison of void and solidness, and inter-catching volumes, displaying the dynamics of jumping fish. The interwoven corridors in melody-like rhythm facilitate the circulation, bringing pleasant experiences to citizens.

The building is enveloped by the latest system and skin material to present the ever changing roof and wall forms in a more modern way and vividly display their natural flowability. The solid part of the building skin is made of quality white acrylic external wall material which allows the integration of LED backlight. The white translucent material, coupled with certain level of reflectiveness and LED application, bestows a mysterious atmosphere to the building. For the translucent part, the irregular translucent frit glass is attached to the elegant spatial frame structure of the building, covering the square and providing a unique daylighting medium for the F&B on terrace.

1 总平面图
  Site plan

2 双鱼夜景,与明华轮交相辉映
  Night view of the double-fish pattern and Minghua Ship

3 秉承招商局"双鱼"旗帜精神,完美连接蛇口地标
  Connection with the landmark in Shekou is realized by carrying on the double-fish spirit of the China Merchants Group

4 以明华轮为中心,双鱼呈环抱姿态,形式优美
  Double fish pattern embraces Minghua Ship at the center elegantly

5/6 露天吧
  Open-air bar

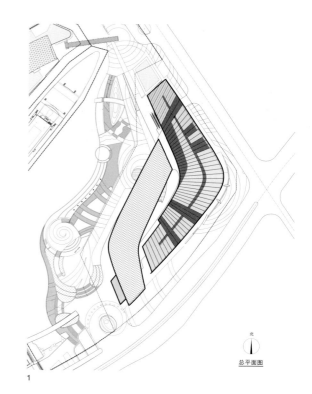

1 总平面图

2

3

4

5

6

Commerce, Office and Hospitality Buildings 商业、办公及酒店建筑

7

8

9

7/8 材料营造的多维空间
Multi-level spaces created by materials

9 连续转变自然流动的墙面屋面形态
Ever changing and naturally flowing wall and roof form

10 立面图
Elevation

11 半透明材质附着于优雅的空间结构，灯光赋予建筑神秘感与未来感
Translucent materials are attached to elegant spatial structures and the lighting adds mysterious and futuristic atmosphere to the building

12/13 绚丽的灯光效果
Gorgeous lighting effect

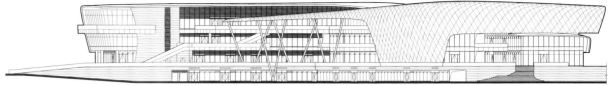
10

11

12

13

Commerce, Office and Hospitality Buildings 商业、办公及酒店建筑 **121**

海上世界广场-船前船尾广场
Seaworld Plaza - Ship-front Square and Ship-end Square

海上世界广场——船前广场、船尾广场位于深圳蛇口国际滨海休闲片区，处于已形成并在未来不断完善的"山海间绿色通道"上的重要节点，是著名的历史航海古港遗迹与城市未来相连接的交点。设计围绕场地东南侧 "海上世界"明华轮这一主题核心展开，融汇明华轮独特的历史文化，展示"海洋"、"运航"等元素，营建出独特的岭南聚落风格的体验型商业街区。

各栋建筑围绕"明华轮"及中央湖面为景观视线焦点布局，利用各组建筑屋面、露台构建成层叠退台的步行休闲系统。组团建筑内设有公共使用的开敞连廊，各组团之间既相对独立，又通过各层连廊使其具有内在的联系，形成一条活跃的主题型步行街动线，有机联系各种组团式商业业态，所形成街区空间连接都市空间，为魅力蛇口添一道靓丽的风景线。退台式商业设计手法，结合景观和商业，面向中央湖面，为商户保留了更多外摆空间，更保证每层平台均可观赏水秀奇观，充分让人们感受景观所带来的愉悦休闲氛围，在层叠的波浪和蜿蜒的小道，让人们置身探索之旅。

对于古港遗迹而言，最具代表性的符号是码头，设计把码头的元素融入到了整个项目里面，以"波浪"为设计主题，在造型设计上充分体现蛇口的开拓精神和当地的海洋文化，与周边现有建筑、人文风格相协调，把岭南建筑上常见的木作、砖墙元素运用到商场的立面当中，重系岭南文化。街区城市设计手法的运用，更进一步把景观中的庭、水、风、光等元素巧妙编织进建筑组团之中，在活跃的形体中透出亲切和清雅。

Located at the important node of the constantly upgraded "green passage between mountains and the sea" in the international coastal leisure district, Shekou, Shenzhen, Seaworld Plaza–Ship-front Square and Ship-end Square represents a connection between the well-known relics of the historical voyage port and the urban future. Centering around the core theme of Seaworld Minghua Ship in the southeast of the site, the design incorporates the unique history and culture of Minghua Ship, displays such elements as sea and navigation, and creates an experience-based commercial neighborhood with a unique Lingnan settlement style.

Buildings are distributed around Minghua Ship and the central lake surface as the visual focus of landscape. Roofs and terraces of building groups constitute a cascading and terraced pedestrian leisure system. Public open corridors are provided inside the building clusters. Relatively independent clusters are internally connected via corridors at different floors to form an active themed pedestrian street circulation, organically connecting various business clusters. The neighborhood spaces thus created connect with urban context, adding another attractive urban space in Shekou. The terraced commercial functions are conceived in view of landscape, facing the central lake with more outdoor business space spared for tenants. This design approach ensures view from each platform to the spectacular water features, making visitors fully immersed in the pleasant leisure atmosphere created by the attractive landscape and encouraging them to explore the winding paths and sea coast.

The most representative element of the ancient port, the wharf, is incorporated into the whole project. With the sea waves as the design concept, the building form fully presents the pioneering spirit and local marine culture of Shekou while harmonizing with the surrounding urban and cultural context. The facade of the shopping mall references the woodwork and brick wall, the typical elements of the Lingnan architecture, to pay tribute to the Lingnan culture. The urban design approach used in neighborhood design further integrate the landscaping elements like courtyard, water, wind and daylight into the building cluster, adding more friendly and elegant experiences to the dynamic building.

项目地点：深圳市南山区蛇口海上世界
设计时间：2010-2013
建设时间：2011-2013
建筑面积：56753m²
建筑层数：3层
建筑高度：16m
合作单位：美国凯里森建筑设计事务所
曾获奖项：深圳市第十六届优秀工程勘察设计（公建建筑）一等奖

Location:	Sea World, Shekou, Nanshan District, Shenzhen
Design:	2010-2013
Construction:	2011-2013
GFA:	56,753m²
Floors:	3
Height:	16m
Partner:	CALLISON
Award:	The First Prize of the 16th Excellent Engineering Survey and Design Award (Public Building) of Shenzhen

1 船前总平面图
  Site plan of Ship-front Square
2 船尾总平面图
  Site plan of Ship-end Square
3 中心广场
  Central square
4 商场入口
  Mall entrance
5 鸟瞰效果图
  Bird's eye view
6 船尾立面
  Elevation of Ship-end Square

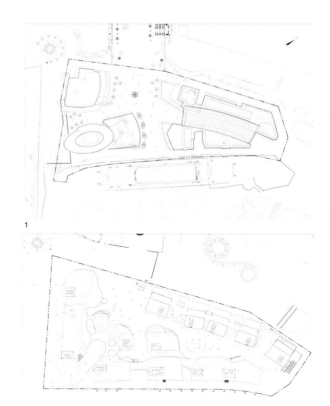

3

4

5

6

昆明西山万达广场—双塔
Xishan Wanda Plaza - Super High-rise Office (Twin Towers)

昆明西山万达广场是万达第100座万达广场，位于云南省昆明市，是万达集团新一代城市综合体项目，包含写字楼、公寓、购物中心、五星级酒店等功能单元。双塔首层为层高10米的大堂，其中10层、22层、34层、46层及58层为避难层（兼作设备层、结构加强层），南塔顶层规划了一个云中会所，其他为层高4.1米的办公标准层。双塔分南塔和北塔，建筑高度为307米（其中南塔地上66层，屋面标高297.3米；北塔地上67层，屋面标高296.6米）。项目从设计到竣工投入使用历时3年，是目前云贵地区的第一高楼，也是国内第一例超300米高的双子塔超高层建筑。

双塔外形从下向上先每层变大，后又渐渐变小，立面为多个双曲面且层层向内退。昆明市花是山茶花，项目整体立面设计灵感便是来自于它。建筑裙房造型以流畅的弧线围合，以石材与玻璃两材质相互穿插，形成不同层次的肌理，富有一片片茶叶的符号。同时结合肌理变化的弧线上设计种植植物，形成一片片绿色的叶子。双塔采用简洁的建筑手法，采用全玻璃幕墙的建筑立面，四向立面及四个建筑转角在简洁的玻璃幕墙上运用幕墙进退的关系营造出山茶花花瓣造型的特点。

昆明西山万达广场（双塔）荣获第十二届（2015～2016年度）第一批中国钢结构金奖，是在8度地震设防烈度区，面对超深基坑、超软地基、超难结构、超高建筑等复杂情况下建设完成的精品建筑。双塔属于超B级高度超限结构，采用了钢管混凝土柱钢梁框架–型钢混凝土核心筒体系，设置钢结构伸臂腰桁架加强层。双塔基坑深达16米，其中的坑中坑则深达23米，超出云南省规范深度8米，创下云南省最深基坑纪录。双塔两块4米厚基础筏板60小时不间断浇筑完成，也创下了昆明市最大体积一次性浇筑量记录。

项目地点：云南省昆明市
设计时间：2012
建设时间：2012.12-2016.12
建筑面积：459600m²
建筑层数：67层
建筑高度：297.3m
获得奖项：2017年度广东省优秀工程勘察设计二等奖

LLocation: Kunming, Yunnan Province
Design: 2012
Construction: 2012.12-2016.12
GFA: 459,600m²
Floors: 67
Height: 297.3m
Award: The Second Prize of Excellent Engineering Exploration and Design Award of Guangdong Province, 2017

Located in Kunming, Yunnan Province, this one-hundredth Wanda Plaza developed by Wanda Group is designed as a new generation of urban complex integrating the functions of office, apartment, shopping mall and 5-star hotel. The ground floor of the twin towers houses 10m-high lobby, while F10/22/34/46/58 serve as the refuge floor (concurrently as mechanical floor and structural reinforced floor). The south tower provides 4.1m-high typical office floor save the sky club on the top floor. The 307m-high twin towers includes the south tower with 66 aboveground floors and roof elevation of 297.3m and north tower with 67 aboveground floors and roof elevation of 296.6m. It took three years to design and complete the highest building in provinces of Yunnan and Guizhou and the first high-rise twin towers over 300m in China.

The building form reversely tapers from the base and then tapers toward the top. The façade is composed by multiple hyperboloids and set back floor by floor. The overall façade design is inspired by the camellia, the city flower of Kunming. The podium is clad with smoothly-curved stone and glaze envelop that alternate with each other, presenting various layers and symbolizing the layers of leaves. Meanwhile, the plants are grown along the curves that follow the change with the texture and outline the green leaves. The twin towers are fully glazed to reflect the concise and clear-cut architectural approach. With the recessed and protruding parts of the façade, the distinctive pattern of camellia petal is created on four facades and four building corners.

As the one of the first batch of winners of the Twelfth Gold Award of China's Construction Engineering Steel Structure (2015-2016), Xishan Wanda Plaza (twin towers) is a super high-rise constructed in a Seismic Intensity Grade VIII region and faced with the challenges of super deep foundation pit, soft foundation and complicated structure. The concrete-filled steel column and steel beam frame – steel-reinforced concrete core combined structure system is adopted with reinforcement floor of steel structure outrigger and belt truss to address over-the-code structure with super Grade B height. The foundation pit is 16m deep and the pit in foundation pit is 23m deep, 8m more than the depth as stipulated in the provincial code, making it the deepest one in Yunnan Province. The two pieces of 4m-thick raft foundation are completed one-time through 60h continuous pouring, the record volume for one-time pouring in Kunming.

1

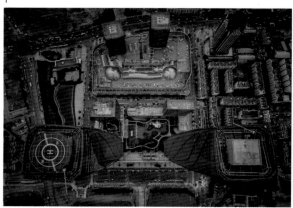

2

3 夕阳下的双塔
Twin towers in sunset

4 繁华永不落幕
Spectacular night view

1 总平面图
　Site plan

2 俯视大地
　Soaring Towers

3 夕阳下的双塔
　Twin towers in sunset

4 繁华永不落幕
　Spectacular night view

5 高原上的双塔
　Twin towers in plateau

Commerce, Office and Hospitality Buildings 商业、办公及酒店建筑 125

6 挺拔清秀
  Aerial renderings

7 旋转的天空
  Rising towers against blue sky

8/9 开敞、大气的大堂
    Generous lobby

Commerce, Office and Hospitality Buildings 商业、办公及酒店建筑 **127**

珠江新城F2-4地块项目
Plot F2-4, Zhujiang New Town, Guangzhou

在广州市天河区珠江新城CBD核心地段，中央广场的绿核公园，是广州标志性的城市核心。珠江新城F2-4地块项目位于其东侧，是高德置地的"冬广场"，与其西侧的"春广场"、"夏广场"、"秋广场"共同构成一个完整的商业群体。

项目主要由三个塔楼及裙楼部分组成，外观造型简洁大方，俯瞰珠江新城中心轴公园。错落的塔楼能够最大限度地扩大各个塔楼的视野，让三个塔楼从不同角度享受珠江新城中心轴公园的城市绿化。

项目拥有大型宴会厅，30米高的顶层酒店大堂及11米高首层多功能大堂。首层设置架空走廊，营造商业灰空间。酒店在高层区域设置多个不同类型餐饮、酒吧，采用错落空间的方式展示其独特魅力。商业部分结合流线设置了错位中庭空间。负一层大中庭的商业展览空间与大阶梯成就独特的舞台视觉效果。裙楼5层开始设置不同空间绿化平台，与屋顶花园形成错落有致的绿化景观，勾勒出别具一格的休闲空间。从花城广场仰望，充满雕塑感的建筑形象闪耀着广州的客厅——珠江新城。

The project is located near the central green at the core of Zhujiang New Town CBD, Tianhe District, a representative urban core of the city. The project on the east, also known as GT Land Winter Plaza, joins its counterparts on the west, including Spring Plaza, Summer Plaza and Autumn Plaza, to form a complete commercial cluster.

Composed of three towers and one podium, the building looks concise and elegant, enjoying a panoramic view of the central green in Zhujiang New Town. The staggered layout of towers can maximize views from each building, having all the three towers overlook, from different angles, urban greening of the central green.

The project features large ballrooms, 30m hotel lobby on top floor and 11-meter functional lobbies on the first floor. Elevated corridors on the first floor offers gray spaces for retail. Hotel F&B and bars are elegantly staggered and distributed on higher floors. Staggered atria spaces are also provided for retail to facilitate the circulation. The commercial exhibition space works with the generous stairs in the major atrium on B1 to present an unique stage-like visual effect. Various green terraces are provided from F5 of podium, and, together with the rooftop gardens, offer attractive views and leisure spaces. When looked up from the Huacheng Square, this sculptural building adds another highlight to Zhujiang New Town, the city's showcase.

项目地点：广州市珠江新城珠江东路
设计时间：2007.9-2013.12
建设时间：2009.9-2015.12
建筑面积：391970.20m²
建筑层数：地下6层，地上分别16F/46F/49F
建筑高度：64.9m/200m/282.8m
合作单位：Woods bagot asia limited 贝格亚洲有限公司
获得奖项：2016年广东省土木建筑学会科学技术一等奖
2013年度广东省优秀工程勘察设计BIM专项三等奖
2014年度广东省优秀工程勘察设计奖专项建筑结构类一等奖
2016年度广东省优秀工程勘察设计奖建筑环境与设备专项一等奖
2017年度广东省优秀工程勘察设计一等奖

Location: Zhu Jiang Dong Lu, Zhujiang New Town, Guangzhou
Design: 2007.9-2013.12
Construction: 2009.9-2015.12
GFA: 391,970.20m²
Floors: 6 underground, 16/46/49 aboveground
Height: 64.9m/200m/282.8m
Partner: Woods Bagot Asia Limited
Awards: The First Prize of Science and Technology Awards of the Civil Engineering and Architectural Society of Guangdong Province, 2016
The Third Prize of BIM Engineering under Excellent Engineering Exploration and Design Award of Guangdong Province, 2013
The First Prize of Building Structure under Excellent Engineering Exploration and Design Award, 2014
The First Prize of Built Environment and Services under Excellent Engineering Exploration and Design Award of GDAD, 2016
The First Prize of Excellent Engineering Exploration and Design Award of Guangdong Province, 2017

1 总平面图
　Site plan

2 黄昏
　Dusk view

3 从花城广场仰望，充满雕塑感的建筑形象闪耀着广州的客厅——珠江新城
　When looked up from the Huacheng Square, the sculpture-like building adds glamour to the city's showcase, the Zhujiang New Town

4
5

6

4-6 立面局部肌理
    Partial texture of facade

7 裙楼首层平面
  Podium F1 floor plan

8 酒店大堂
  Hotel lobby

9 中庭
  Atrium

10 办公楼大堂
   Office lobby

7

130 商业、办公及酒店建筑 Commerce, Office and Hospitality Buildings

Commerce, Office and Hospitality Buildings 商业、办公及酒店建筑

保利商务中心
Poly Business Center

保利商务中心是集商业、办公、星级酒店于一身的大型综合性项目，是佛山东平新城内地标性建筑之一。

建筑通过石材与玻璃强烈的虚实对比产生刚劲有力的美感，美观的镂空金属板和浅色天然石材的结合，展现了商业中心主体建筑的典雅高贵形象。塔楼整体设计创意，采纳了中国传统习俗中的天灯。在超过200米的高度上的柔性玻璃幕墙系统（索网幕墙），完美地展现"天灯"的造型，双曲面索网结构表面覆盖张拉膜，创造出丰富多变的建筑造型，充分展示出膜结构特有的艺术感染力。

办公塔楼高248米，塔楼顶部220米以上为标志性的超高空天灯造型，29.7m高的天灯层四周为大面积单层索网幕墙。结构设计针对性的采用了适合建筑功能分区要求的结构形式，整合了钢筋混凝土结构、钢-混凝土混合结构以及钢结构等多种类型而成为特殊的混合结构体系。天灯层位于塔楼顶部5层，结构设计采用空间钢框架+单层双向平面索网幕墙结构体系。在建筑的四个角部设置了四根格构柱，格构柱的顶部由水平钢桁架拉结成为稳定的空间钢框架体系。在钢框架围闭的区域内，设置了四片单层双向平面索网幕墙。

As a large mix-used project integrating retails, office and starred hotel, Poly Business Center represents one of the landmarks of Dongping New Town in Foshan.

The sharp contrast between the solid stone and transparent glass reveals the strength and aesthetics, while the combination of the artistic hollow-out metal plate and natural stones in light color presents an elegant and noble image of the main building of the business center. Inspired by the "sky lantern" from the traditional Chinese customs, the overall concept is to present a sky lantern on the flexible glass curtain wall system (cable net curtain wall) at a height above 200m. The double curved surface of the cable net structure covered with tensile membrane diversify the building form and give full play to the unique artistic attraction of the membrane structure.

Embedded into the top of the 248m office tower, the 29.7m iconic "sky lantern" rises from above 220m and is wrapped with large expanses of single-skin cable net curtain wall. The structure is specifically designed to accommodate the building's functional zoning, integrating reinforced concrete structure, steel-concrete composite structure and steel structure etc. into a special composite structural system. The sky lantern occupies the top five floors of the tower with the structure system of spatial steel frame + single-layer two-way horizontal cable net curtain wall. Four lattice columns are provided at the four corners of the building with horizontal steel truss at the top tied to into a stable spatial steel frame system. Four pieces of single-layer two-way horizontal cable net curtain wall are provided for the area enclosed by the steel frame.

项目地点：广东省佛山市东平新城文华路
设计时间：2010-2012
建设时间：2012
建筑面积：20.7万m²
建筑层数：地下2层，地上22层（酒店塔楼）；地上54层（办公塔楼）
建筑高度：248m（办公塔楼）；99m（酒店塔楼）
合作单位：gmp International GmbH

Location:	Wen Hua Lu, Dongping New Town, Foshan, Guangdong Province
Design:	2010-2012
Construction:	2012
GFA:	207,000 m²
Floors:	2 underground and 22 aboveground (hotel), 54 aboveground (office)
Height:	248m (office); 99m (hotel)
Partner:	gmp International GmbH

1

2

3

4

5

1 鸟瞰图
  A Bird's eye view

2 总平面图
  Site plan

3 隔江远眺东平新城的美景
  Attractive view of Dongping New Town across the river

4 华灯初上，熠熠生辉
  Sparkling in the city's nightscape

5 镂空金属板与三角形石材幕墙单元形成的肌理效果
  Fabric effect produced by hollow-out metal plate and triangular stone façade unit

Commerce, Office and Hospitality Buildings 商业、办公及酒店建筑

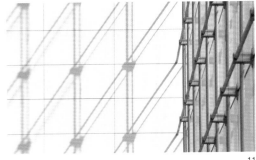

6/7 镂空金属板的细节设计
Detail design of hollow-out metal plate

8 别具心思的外墙有如凝固的音乐旋律
Ingeniously designed exterior wall, like frozen melody

9 "天灯"构思与实景
The concept and view of "sky lantern"

10/11 位置最高的平面索网幕墙实景
View of the horizontal cable-net curtain wall at the top

交通建筑

在广东省建筑设计研究院60多年的历史中，广州新白云国际机场一期航站楼项目无疑是有着广泛社会影响力的作品，它是当年中国民航史上一次性投资最大的项目之一，是广东省当年规模最大、难度最高、功能复杂、设计新颖、技术先进的公共建筑工程项目，它代表着广东省建筑业当年最为卓越的水平，也是中国民航史上一次转场成功的特大型航站楼的范例，展示了新世纪中国大型标志性建筑的先进性与独特性。作为广州市的重要门户之一，广州新白云国际机场为南来北往的旅客提供了优质的服务。

在1998年初，白云机场业主为新白云机场举行国际竞赛，包括英国福斯特事务所与荷兰机场公司（NACO）联合体、法国巴黎机场公司（ADP）、美国PARSONS公司与美国URS Greiner公司联合体、加拿大B+H建筑事务所等全球知名机场设计公司参加。GDAD的领导班子毅然决定组织技术骨干（即机场组）第一次参加重大国际竞赛，认为这是GDAD一个千载难逢的好机会。通过参加竞赛拓展视野，向国际公司学习，更为重要的是展现我们的综合实力，争取做后续的合作设计。在各专业院老总的指导下，机场组经过三个月的努力拼搏，终于交出了一份令评委和业主都留下深刻印象的竞赛方案。在1998年的金秋十月，GDAD被白云机场业主确定为新机场的中方设计单位，与美国公司合作设计。从此，一个光荣的使命等待GDAD去完成！

新白云机场一期工程来之不易，全院从多个部门抽调技术骨干70多人组成新的机场设计组，包括建筑、结构、给排水、电气、空调、市政、交通和概预算全部专业。2004年，他们经过1600多个日夜鏖战，完成广州白云国际机场一号航站楼这项世纪工程，擦亮中国的南大门，经过这次的中外合作设计，先后掌握了航站楼工艺（航站楼构型、旅客流程、行李流程、联检流程和信息流程）、大跨度建筑的外围护结构技术（大型金属屋面系统、大面积点式玻璃幕墙系统、张拉膜系统）、大跨度结构技术（桁架技术和无檩式箱形压型钢板、三管梭形钢格构人字柱）、异常复杂的石灰岩岩溶发育地区的基础技术，大型公共建筑的能源管理系统（EMS）、高大空间空调技术、消防设计等关键技术，积累了丰富的大型公共建筑设计经验，尤其在大跨度大空间建筑设计的各个相关专业，GDAD完成原始积累而进入了全国先进行列。而且，我们不单在技术上有所突破，在大项目管理及现场服务方面也积累了丰富的经验，为承接大型复杂公共建筑设计奠定了技术、管理和服务的坚实基础。

继新白云机场之后，GDAD先后参加了其他机场、火车站、体育馆的竞赛和投标，如昆明机场、长沙机场、南宁机场、潮汕机场、重庆机场、深圳机场、武汉机场、浦东机场等十多个机场的投标，参加了亚运会的几个场馆竞赛，继续在大跨度建筑方面探索创新，并从机场航站楼延伸到体育馆、火车站和交通枢纽等，陆续完成了一批有影响力的作品。其中比较突出的有新白云机场T1航站楼扩建、潮汕机场、广州亚运馆、广州亚运自行车馆、花都亚运馆、惠州游泳跳水馆、武汉火车站（武广高铁）、昆明南高铁站、深圳机场航站楼交通中心、广州地铁1号线、2号线等一批轨道站。最近，GDAD参加国际招标并中标了深圳机场卫星厅设计。

目前，GDAD负责设计的新白云机场T2航站楼已于2013年动工，计划2018年初投入使用。目标是打造亚太地区连接各大洲新的航空枢纽，朝着世界级机场迈进！

Transportation Buildings

In GDAD's over 60 years of history, Guangzhou Baiyun International Airport Phase I Terminal is undoubtedly a project enjoying extensive public reputation. As one of the most costly one-off investment projects in China's civil aviation history, it dwarfed other public buildings in Guangdong in terms of size, difficulty, functional complexity, design innovation and technical sophistication, representing the highest level of building industry of Guangdong at that time. It also sets up a good example of one-time relocation of super large terminals in China's civil aviation history, showcasing its leading position as a unique landmark in China in the new century. Being one of the key gateways of Guangzhou, the new Guangzhou Baiyun International Airport offers quality services to the domestic and international passengers.

In early 1998, an international competition was launched for the new Baiyun International Airport, attracting prestigious airport designers like Foster + Partners + NACO (consortium), ADP, PARSONS + URS Greiner (consortium) and B+H etc. GDAD's leadership decided to mobilize a capable team (i.e. the Airport Design Group) to participate in the international competitor for the first time. The competition was regarded as a great opportunity to expand our horizon, learn from international companies and more importantly to show our well-rounded competence and win the contract for the subsequent design tasks. Under the direction of principals of various divisions, the Airport Design Group finally presented a design proposal that greatly impressed the jury and the client after three months of hard work. In the autumn of 1998, GDAD was designated as the local design institute to complete the airport design in cooperation with an America-based company.

For the hard-won Project, GDAD mobilized more than 70 capable technical personnel from various divisions to set up a full-specialty airport design team, including architecture, structure, water supply, electrical, AC, utility, transportation and budgetary estimate. In 2004, after 1,600 days of indefatigable hard work, GDAD completed the design of this centurial project, polishing up the image of the city known as the China's south gate. Through the international cooperation on this project, GDAD acquired the key technologies such as terminal design (terminal configuration, passenger flow, baggage flow, joint inspection flow and information flow), envelope structure of large span building (large metallic roof system, extensive point-type glass curtain wall system, tensile membrane system), large span structure (truss and non-purlin box-profiled steel sheet, 3-pipe shuttle-shape steel latticed column), foundation in extremely complicated karst developed area, energy management system (EMS) of large public building, as well as AC, fire and other key technologies of tall spaces. Having accumulated considerable design experiences of large public building, in particular those relating to large-span and large-space architectural design, GDAD became one of the leading design institutes in China through this important practice. In addition to the technical breakthroughs, GDAD also developed expertise in large project management and construction administration, making solid inroads into large complicated public building design in terms of technology, management and services.

Right after the new Baiyun International Airport, GDAD participated in the design competitions for other airports, railway stations and gymnasiums, including Kunming Airport, Changsha Airport, Nanning Airport, Chaoshan Airport, Chongqing Airport, Shenzhen Airport, Wuhan Airport and Pudong Airport. We also participated in the competition for several sports venues of Guangzhou Asian Games to keep exploring the innovation in large-span building, extending the building typology from the airport terminal to gymnasium, railway station and transportation hub. The representative projects include Baiyun International Airport T1 Expansion, Chaoshan Airport, Guangzhou Asian Games Gymnasium, Guangzhou Asian Games Velodrome, Huadu Asian Games Gymnasium, Huizhou Swimming and Diving Natatorium, Wuhan Railway Station (Wuhan-Guangzhou High-speed Rail), Kunming South High-speed Railway Station, Shenzhen Airport Terminal Transportation Center and metro stations along Guangzhou's Metro Line 1 and 2. Recently, GDAD won the international design competition for satellite hall of Shenzhen Airport.

Currently, GDAD-designed Baiyun International Airport T2 is under construction, which was commenced in 2013, and will be put into service in early 2018. The project goal is to create a new aviation hub and a world-class airport that connects the Asia-Pacific region with other continents.

广州新白云国际机场一号航站楼
Terminal 1, Guangzhou New Baiyun International Airport

夜幕降临,随着一架架飞机降落在新机场,随着老机场跑道灯光的渐渐熄灭,一个令世人瞩目的时刻到来了:2004年8月5日,广州新白云国际机场正式启用。作为我国三大枢纽机场之一,一个按照中枢理念设计、建设、营运的崭新机场,经过近10年的筹划,4年的规划设计,3年零10个月的建设,终于在花都一片宽阔的土地上崛起。航站楼造型新颖,具有明显的中轴线,展现了强烈的标志性。在蓝天白云之下,地平线上一组流畅有力的弧线,勾勒出独特的建筑形象,高低起伏,充满动感,在亚热带植物的衬托下体现出中国南大门的雄伟气势。新机场既是广东建筑业的一大成就,也是我国民航史上一个重要的里程碑,反映了中国特别是广州最新的建筑技术水平,展示了新世纪中国大型标志性建筑的先进性与独特性。

航站楼设计年旅客吞吐量2500万人次,高峰小时客运量9300人。航站楼构形在国内外机场中独一无二,采用"主楼+双向连接楼+指廊式"构形。航站楼由主楼、东连接楼、西连接楼和四条指廊组成,拥有总计66个机位(近机位46个,远机位20个)。其最大特点是出港旅客在主楼三层办理登机手续,在指廊分散候机和登机,到港旅客由指廊分别经东、西连接楼首层提取行李离开。主楼具有可从南、北方向双向进入的特点,具有双倍的车道边。航站楼扩建时向北发展,不影响首期营运。这是国内第一个将地铁站设在出港大厅下面的航站楼,是机场和地铁紧密结合的杰出工程范例。

本项目在中国首次大面积使用预应力自平衡索结构点支式玻璃幕墙,张拉膜用于覆盖复杂的曲面造型,也是在中国机场中第一次使用。这还是中国目前在岩溶地区兴建的最大规模的民用建筑项目,在中国首次使用三管梭形钢格构人字柱,以及大跨度屋面无檩式箱形压型钢板。

项目地点:广州市北部,白云区人和镇与花都区新华镇的交界处
设计时间:1998-2002
建设时间:2000-2004
建筑面积:353042m²
建筑层数:地上3层,地下2层
建筑高度:55.88m
合作单位:美国PARSONS公司 + URS Greiner公司
曾获奖项:2011年评为中国"百年百项杰出土木工程"
2010年中国建筑学会建筑创作大奖
2007年度广东省优秀工程技术创新奖
2006年度"全国优秀工程勘察设计金质奖"
2005年获得第五届"詹天佑土木工程大奖"
2005年"全国十大建设科技成就"称号
2005年获得"首届全国绿色建筑创新奖"
2005年度广东省优秀工程勘察设计奖一等奖

When the runway lights of the old airport went out at nightfall and the aircrafts started to land at the new airport, Aug. 5, 2004 marked a tremendous moment when Guangzhou New Baiyun International Airport, one of China's three major hub airports, was officially put into operation. The new hub airport, after almost 10 years of preparation, 4 years of design and 3 years plus 10 months of construction, finally rises in Huadu District, Guangzhou. The highly representative terminal building features novel shape and a distinct central axis. The sweeping arc lines undulate dynamically to portray a unique terminal image against blue sky and white clouds, and, coupled with the subtropical plants, highlight the grandeur of this South Gateway to China. The Project is not only a great achievement in Guangdong's building industry but also a key milestone in China's civil aviation history. It showcases the latest building technology of China particularly Guangzhou and the advancement and uniqueness of large landmark buildings in China in the new century.

Designed with a passenger traffic of 25 million, the terminal receives 9,300 passengers at the peak hour. The terminal configuration comprising main building, two-way connecting buildings and piers is the first of its kind in the world. Specifically, the terminal includes main building, east connecting building, west connecting building and four piers, offering 66 stands in total, of which 46 are near stands and 20 are remote stands. The departing passengers may check in on F3 of the main building before arriving at different piers for waiting and boarding. Arriving passengers may claim their luggage on the ground floor of the main building after exiting the piers and passing the east/west connecting building. With two curbsides, the main building can be accessed from both the north and the south. The terminal is planned to expand northward to avoid impact on the operation of Phase I. As the first terminal in China which places a metro station below the departure concourse, it sets up a noticeable example of airport-metro integration.

The Project is the first in China to extensively use the point-supported glass curtain wall with pre-stressed self-balance cable structure, and the first in Chinese airport to employ tension membranes to cover sophisticated curved shape. It is also the largest civil building ever built in China's karst area by now, and the first Chinese project that uses three-tube fusiform steel latticed herringbone column and large-span purlin-free roof box profiled steel sheet.

Location: Crossing of Renhe Town, Baiyun District and Xinhua Town, Huadu District in the north of Guangzhou
Design: 1998-2002
Construction: 2000-2004
GFA: 353,042m²
Floors: 3 aboveground, 2 underground
Height: 55.88m
Partner: PARSONS + URS Greiner
Awards: "100 Outstanding Civil Engineering Projects from 1900 to 2010", 2011
Architectural Creation Award by the Architectural Society of China, 2010
Excellent Engineering Technology Innovation Award of Guangdong Province, 2007
Gold Award of National Excellent Engineering Exploration and Design Award, 2006
The Fifth Tien-Yow Jeme Civil Engineering Prize, 2005
Top-10 National Construction Technology Achievement Award, 2005
The First National Green Building Innovation Award, 2005
The First Prize of Excellent Engineering Design Award of Guangdong Province, 2005

1 白云机场总平面图(含一号、二号航站楼)
　Site Plan of the New Baiyun International Airport (including T1 and T2)
2 主楼出发大厅中央的张拉膜采光带引导旅客分别前往东/西连接楼
　Tensile membrane daylighting band in the center of Main Building's departure hall guides passengers to the east/west connecting building

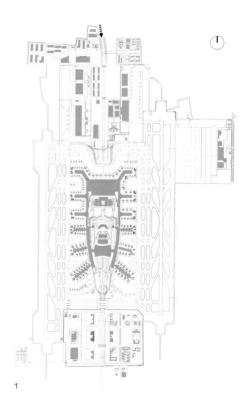

1

3

4

5

8

140　交通建筑 Transportation Buildings

3 "主楼+双向连接楼+指廊式"的独特构型
  Unique configuration comprising Main Building, Two-way Connecting Buildings and Piers

4 东西连接楼夜景:"老虎窗"及张拉膜雨篷
  Night view of east/west connecting building: dormer and tensile membrane canopy

5 绚丽的航站楼夜景:主楼大面积点式玻璃幕墙及张拉膜雨篷
  Gorgeous night view of terminal: extensive point-supported glass curtain wall and tensile membrane canopy of main building

6 连接楼二层到达走廊:弧形点式被动幕墙使空侧机坪景观开阔舒展
  Arrival corridor on F2 of connecting building: arc-shaped point-supported passive curtain wall presents generous view toward the airside apron

7 候机厅的张拉膜采光带:指引旅客流程并避免眩光
  Tensile membrane daylighting band in the waiting hall: guides passenger flow while avoiding glare

8/9 立面图
   Elevation

Transportation Buildings 交通建筑

广州新白云国际机场二号航站楼
Terminal 2, Guangzhou New Baiyun International Airport

广州亚运会以来新白云机场持续快速发展，2016年旅客量超过5500万人次，已超过1998年规划的总量，进入世界机场前15名。因此，T2航站楼及配套设施的建设势在必行。这是目前我国在建的规模最大的航站楼之一，面积超过63万平方米，配套交通中心（GTC）及停车楼21万平方米，设计容量为年旅客量4500万人次，近机位70个。T2航站楼于2013年动工，计划2018年完成施工投入使用。届时，新白云机场设计总年旅客容量将达到8500万人次，以打造亚太地区连接各大洲新的航空枢纽为目标，朝着世界级机场迈进。

T2航站楼在T1航站楼北面进行扩建，保持南北贯通的航站区布局，北进场车流通过主楼下的隧道与南侧陆侧交通路网连接。T2航站楼包括主楼、六条指廊及北指廊，其中东四和西四指廊分期建设，航站楼规划将"分离站坪"改为"北站坪"概念，采用"指廊式＋前列式"混合构型，获得多个紧靠主楼北站坪的前列式大型机位，使国际－国内机位可灵活转换使用，并有效缩短了旅客步行距离。旅出发旅客从南侧车道边进入主楼办票大厅，再前往安检大厅或联检大厅接受检查，最后分流到达候机厅。到达旅客集中在主楼提取行李，主楼首层为国际和国内迎客厅。旅客流程简洁，方向清晰。本期建设预留旅客捷运系统（APM）为出发、到达，特别是中转旅客的优质服务预留的条件。在交通中心（GTC）实现大巴、出租车、社会车、地铁和城轨多种交通方式的快速换乘，人车完全分流。

T2航站楼拥有与T1航站楼和谐一致的建筑造型，保留弧线形的主楼屋面和人字形柱及张拉膜雨篷等特有元素；体现岭南地域特色的花园空间及装修设计；注重绿色环保节能设计，达到中国绿色建筑三星标准；成为展示公共艺术的枢纽门户；反映当今中国最新的建筑技术水平。

项目地点：广州市北部，广州新白云机场航站区
设计时间：2005-2016（2013年动工，计划2018年投入使用）
建设时间：2013年至今
建筑面积：634000m² （2020年，不含东四西四指廊）
建筑层数：地上5层，地下2层（GTC）
建筑高度：44.675m
顾问机构：美国MA公司、美国L&B公司等
曾获奖项：2015年度广东省优秀工程勘察设计奖BIM一等奖

Location: Terminal area of Guangzhou New Baiyun International Airport in the north of Guangzhou
Design: 2005-2016 (Construction commenced in 2013 and to be put into use by 2018)
Construction: 2013 to date
GFA: 634,000m² (by 2020, excluding the eastern Fourth and western Fourth pier)
Floors: 5 aboveground, 2 underground (GTC)
Height: 44.675m
Partner: MA + L&B
Award: The First Prize of BIM under Excellent Engineering Design Award of Guangdong Province, 2015

Guangzhou New Baiyun International Airport has experienced a rapid development since Guangzhou Asian Games in 2010. Its passenger traffic exceeded 55 million in 2016, which already exceeds the total capacity planned in 1998, ranking it among the top 15 airports worldwide. This makes it imperative to develop Terminal 2 and supporting facilities. Terminal 2 is currently the largest terminal under construction in China, with a GFA of over 630,000m², including 210,000m² for a supporting ground transportation center (GTC) and a parking structure. The Project is designed with an annual passenger traffic of 45 million and 70 near stands. The construction commenced in 2013 and will be completed by 2018. By then, with a total annual passenger traffic of 85 million, Guangzhou New Baiyun International Airport will make a big step forward toward its goal to be a new aviation hub in the Asia-Pacific region and a world-class airport that welcomes worldwide passengers.

Terminal 2 is an expansion north of Terminal 1. While a north-south layout is maintained for the terminal area, the vehicles approaching from the north is led to the landside road network in the south through the tunnel under Main Building. Terminal 2 is composed of a main building, six piers and a northern pier, among which East 4th and West 4th piers are developed by phase. The previous "separated apron" concept is replaced with "north apron", so a mixed "pier + linear" stand layout and increased number of linear stands for large aircrafts close to the north apron of Main Building are made possible. These stands can be flexibly used for either international or domestic flights and effectually shorten the walking distance of passengers. Departing passengers will enter the check-in hall in Main Building through the curbside platform in the south, and proceed to Security Hall or Joint Inspection Hall before they are dispersed to gate lounges after inspection. Arriving passengers may gather in Main Building to claim their luggage as International Arrival Hall and Domestic Arrival Hall are both located on the ground floor of Main Building. The circulations for departing and arriving passengers are simple and well-defined. Interface conditions for automated people mover (APM) system are reserved at this phase to provide quality service to departing and arriving passengers and particularly transit passengers in the future. The GTC located south of Main Building can enable rapid interchange between bus, taxi, private car, metro and urban rail transit, thus completely separate pedestrian from vehicular circulation.

Terminal 2 maintains a building shape that is harmonious with Terminal 1, including the arc-shaped Main Building roof, herringbone columns and tensile membrane canopy. It provides garden spaces and fit-out design that embody the local features of Lingnan; conducts green and energy efficient design that is up to China's three-star green building standards; serves as a hub/gateway showcasing public artworks; and reflects the latest building technology of China.

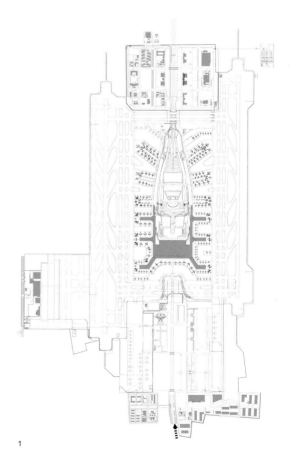

1

1 总平面图
  Site plan

2 T1、T2航站楼全景鸟瞰图
  Master Brid's Eye View of Terminal 1 and 2

3 T2航站楼日景鸟瞰图
  Brid's Eye View of Terminal 2 in Dayligh

4 T2航站楼造型更为简洁流畅,保留了弧形屋面、人字形柱及张拉膜雨篷的设计元素
Concise and smooth shape of T2, continuing the arc-shaped roof, herringbone columns and tensile membrane canopy used in T1

5 航站楼与交通中心连接处效果图
Connection between T2 and GTC

6 陆侧出发层车道边近景效果图
Close-shot rendering of curbside on airside departure level

7

8 9 10

7 具有波浪形吊顶的办票大厅,空间设计简洁流畅,色调明快,外部绿化景观内渗
   Check-in hall with wavy suspended ceiling, concise spatial design, bright color and penetration of exterior green landscape

8 国内候机大厅:候机厅内设有商业,等候空间与商业空间紧密结合,丰富人的活动
   Domestic waiting hall: commercial spaces closely integrated with waiting spaces are provided inside the waiting hall to enrich people's activities

9 国际出发商业空间设有天窗、侧窗,引入花园景色及阳光,使传统的商业空间焕然一新
   Commercial spaces in international departure hall are provided with skylight and side windows to allow in garden view and sunlight which can refresh the traditional commercial atmosphere

10 国际到达走廊采用大面积落地玻璃幕墙,国际旅客到达广州后第一眼就能看到走廊外特具岭南特色的园林景观,充分体现新白云国际机场作为广州门户的作用
   International arrival corridor is provided with extensive floor-to-ceiling glass curtain wall to make the garden landscape of Lingnan characteristics outside the corridor readily visible by arriving international passengers, which fully demonstrates the role of New Baiyun International Airport as the gateway to Guangzhou

揭阳潮汕机场航站楼及配套工程
Jieyang Chaoshan Airport Terminal and Supporting Works

揭阳潮汕机场是省内继广州新白云机场和深圳宝安机场之后的第三大机场，主要服务粤东地区及闽南部分地区。潮汕机场的落成，使该区域摆脱军民合用机场的限制，真正拥有一座现代化的民用干线机场，对于形成粤东地区交通一体化和经济一体化，推动粤东地区经济发展具有十分重要的意义。

潮汕机场航站楼属中型机场规模，将分三期建设。一期航站楼采用了"指廊式+前列式"的复合构型，在相同用地条件下提供最多的近机位，用地效率高。三条与主楼连接的指廊长度约200米，步行距离适中，旅客流程便捷清晰。将来在不停航的前提下，主楼和指廊都可以按需要进行扩建。扩建后仍然是一座具有完整性、一体化的航站楼。航站楼内部空间设计借鉴了传统民居的庭院概念，在主楼和指廊之间设有一个悬空式花园。无论在办票厅、候机厅还是到达走廊、行李提取厅，旅客都可以欣赏到美丽的花园景色，给旅客留下深刻印象。同时，花园还促进航站楼的通风，结合大挑檐、室外遮阳、局部天窗等建筑元素，降低建筑能耗。

潮汕机场于2011年12月建成使用，为来往潮汕地区的旅客提供了优质服务。其正立面优美的轮廓，高低起伏富有韵味。整体造型一气呵成，表达了航空的"飞翔"意念，运用简洁的非线性金属屋面与曲面玻璃幕墙相结合，流畅精致。远期造型犹如飞行器一般，展开双翅，以一种大气、热情的姿态迎接往来旅客，其标志性、时代性和地域特征得到较好的平衡与协调。无论是建筑造型还是空间设计，潮汕机场航站楼都是一个根植于传统地域文化与气候特征，一个探索当代岭南建筑创新设计的原创建筑，一个集全新旅客体验、节能环保、高效运营需求于一体的独特的花园式建筑，其创新设计为我国干线机场航站楼的规划设计提供了有益启示。

Jieyang Chaoshan Airport, as the third major airport in Guangdong following Guangzhou Baiyun International Airport and Shenzhen Bao'an International Airport, mainly caters for the travel demand of east Guangdong Province and south Fujian Province. This modern civil main airport helps East Guangdong shake off the restraint by the former civil and military airport and plays a significant role in realizing regional traffic and economic integration and bolstering economic growth.

The medium-scaled Chaoshan Airport is developed in three phases. The Phase I terminal features a composite structure of "pier+ linear" to maximize the near stands for high efficiency of site. Three 200m-long airside concourses leading to the main building ensure moderate walking distance and easy and clear passenger circulation. Without suspending the service, both the main building and the airside concourses can be expanded in the future as needed with ensured holistic and integrated layout. With reference to the courtyard of traditional house in the design for interior space, a suspended garden is provided between the main building and the airside concourses to impress the passengers with attractive view whether in check-in hall, waiting hall, arrival corridor or baggage claim. In addition, the garden facilitates the ventilation of the terminal and reduces energy consumption of the building with the help of other architectural elements such as large cornice, sunshade and local skylight.

Completed and put into use in Dec. 2011, Chaoshan Airport offers quality service to the passengers traveling inbound and outbound Chaoshan region. The attractive silhouette of front façade and smoothly undulating and stretched building shape reveal the "flying" concept of airlines. The combination of concise non-linear metallic roof and cambered glass curtain wall creates fluid and delicate building appearance. Looking like an aircraft warmly embracing the passengers, the building is well balanced and coordinated in terms of symbolism, contemporaneity and local features. In both building shape and spatial design, the terminal of Chaoshan Airport rooted in local culture and climate represents the original design to explore the innovation of modern Lingnan architecture. The innovative design of this unique garden-like building integrating the innovative passenger experience, energy efficiency and efficient operation also inspires the planning and design of main airport terminal nationwide.

项目地点：广东省揭阳市揭东县炮台镇
设计时间：2007-2010
建设时间：2008-2011
建筑面积：58752m²
建筑层数：地上3层，地下1层
建筑高度：30.17m
合作单位：美国兰德隆布朗公司（规划）
上海新时代机场设计有限公司广州分公司（民航类弱电）
曾获奖项：2013年度全国优秀工程勘察设计行业奖建筑工程三等奖
2015香港建筑师学会两岸四地建筑设计大奖卓越奖
2013年度广东省优秀工程勘察设计奖二等奖

Location:	Paotai Town, Jiedong County, Jieyang City, Guangdong Province
Design:	2007-2010
Construction:	2008-2011
GFA:	58,752m²
Floors:	3 aboveground, 1 underground
Height:	30.17m
Partner:	Landrum & Brown (planning); Shanghai Civil Aviation New Era Airport Design & Research Institute Co., Ltd. Guangzhou Office (ELV design of civil aviation)
Awards:	The Third Prize of Architectural Engineering Design under National Excellent Engineering Exploration and Design Award, 2013; Excellence Award under Cross-Strait Architectural Design Awards by Hong Kong Institute of Architects, 2015; The Second Prize of Excellent Engineering Design Award of Guangdong Province, 2013

1 总平面图
  Site plan

2 花园式航站楼概念草图
  Conceptual sketch for garden-like Terminal

3 航站楼与指廊之间的悬空花园，旅客在出发和到达的流程中可以欣赏到当中的景色，是潮汕机场空间特色之一
  The suspended garden between the Terminal and the airside concourse impresses passengers with attractive views along departure and arrival circulation, as one of spatial features of the airport

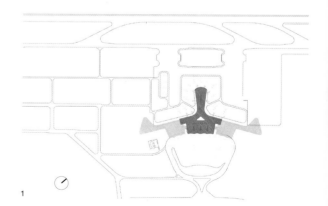

4

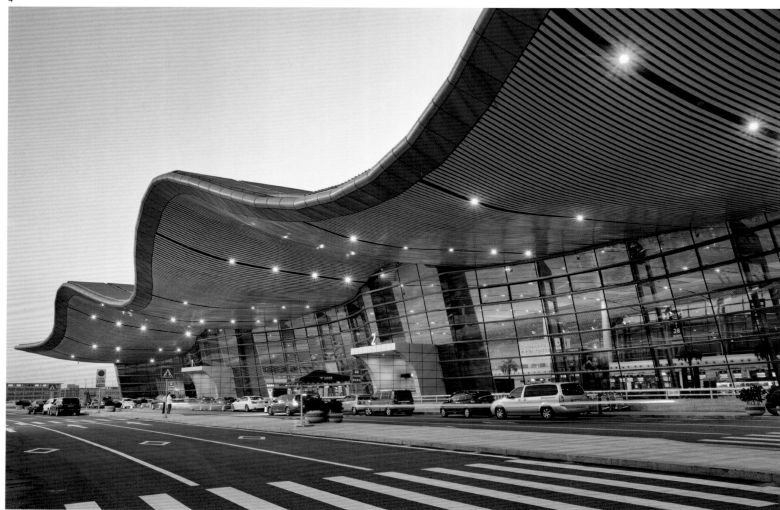

5

4 夜幕下的潮汕机场
  Night view

5 航站楼的造型表达了航空的"飞翔"
  理念，运用简洁的非线性金属屋面与
  曲面玻璃幕墙相结合，流畅精致
  The terminal shape reveals the
  "flying" concept of aviation with fluid
  and delicate appearance created
  by concise non-linear metallic roof
  and cambered glass curtain wall

6 侧立面图
  Side elevation

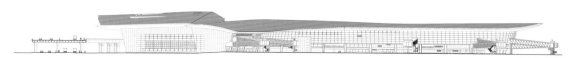

6

148 交通建筑 Transportation Buildings

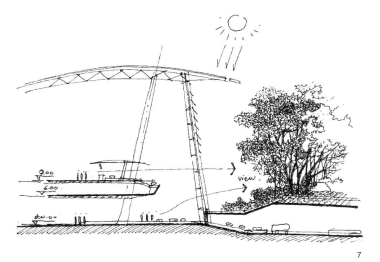

7 观景视线与日光处理手稿
  Sketch of viewing sight line and daylighting concept

8 花园促进航站楼的自然通风，结合飘檐、遮阳等手段降低建筑能耗
  The garden facilitates the natural ventilation, and coupled with the overhung eave and sunshade, cuts the energy demands

9 航站楼剖透视概念图：花园处于航站楼中心
  Conceptual sectional perspective of the Terminal: the garden lies at the center of Terminal

Transportation Buildings 交通建筑　149

深圳机场卫星厅
Shenzhen Airport Satellite Concourse

深圳宝安国际机场是服务珠江三角洲城区三大机场枢纽之一，也是中国第四大及世界第24大繁忙的机场。随着深圳地区航空业客量的迅猛增长，预计到2025年，机场年旅客吞吐量将达到5200万人次。为保证服务质量，将建设机场卫星厅。广东省建筑设计研究院与美国兰德隆布朗公司、香港Aedas组成联合体，赢得卫星厅方案征集竞赛，成为中标单位。

深圳机场卫星厅项目建设面积约为24万平方米，含71个泊位，年客运吞吐量为2200万，高峰小时旅客吞吐量4490人，将为潜在商业发展提供额外空侧空间。

本次卫星厅建筑设计遵循三大核心设计原则：1、高效灵活；卫星厅整体南移，避开地铁风井，重新优化原方案X构型，减少了占地面积，加大港湾夹角，平顺岸线，提高机坪效率。在航站楼与卫星厅之间捷运及行李系统设计考虑周全，提供最大的灵活性，两套系统均能达到投资及旅客服务水平的完美平行。2、人性关怀；卫星厅内以流程优先，步行距离短，旅客流程简洁，指廊空间导向清晰，中转流程所需时间满足IATA对应服务MCT标准。3、卓越成功；卫星厅建筑造型采用标准化的构件建造，简洁的造型和高效的结构能降低运营风险，满足绿色节能建筑标准。多元化的商业布局空间与非航运营策划理念，提升机场非航空赢利能力，创造更多收益。

项目地点：深圳市宝安大道
设计时间：2016.6
建筑面积：240000m²
建筑层数：地下1层，地上3层
建筑高度：29m
合作单位：兰德隆与布朗、凯达环球建筑设计咨询有限公司
曾获奖项：2016年方案国际招标第一名

Location: Bao An Da Dao, Shenzhen
Design: 2016.6
GFA: 240,000m²
Floors: 1 underground, 3 aboveground
Height: 29m
Partner: Landrum & Brown, Aedas
Award: The First Place of International Bidding 2016

Shenzhen International Airport is one of the three air hubs serving the Pearl River Delta, boasting the fourth major airport in China and the 24th busiest airport in the world. With the soaring air traffic of Shenzhen, the airport's annual passenger traffic is expected to reach 52 million by 2025. To ensure the quality service, a satellite concourse was planned for which a design competition was launched and won by the consortium of GDAD, Landrum & Brown and Aedas.

The Project is planned with a GFA of 240,000m², 71 parking spaces, an annual passenger traffic of 22 million and 4,490 per peak hour. Besides, it will offer more airside space for the potential commercial development.

The architectural design for the project follows three design guidelines: 1. Being efficient and flexible. The whole satellite concourse is moved southwards to avoid the metro ventilation shaft. The original X-shaped configuration is optimized to reduce the footprint area, increase the included angle of harbor, smooth the shoreline and enhance the efficiency of apron. The people mover and baggage handling system between terminal and satellite concourse are thoughtfully designed with maximum flexibility to ensure the desirable balance between cost and service quality. 2.Being human-oriented. The design for satellite concourse makes the circulation a top priority to realize short walking distance, streamlined passenger circulation, clear orientation in concourse and transit time up to IATA's MCT standard. 3.Successful business operation. Standardized components, concise form and efficient structure help reduce the operation risk and meet the standards for green and energy efficient building. The diversified commercial space layout and non-aviation operation concept enhance the airport's non-aviation revenue and profitability.

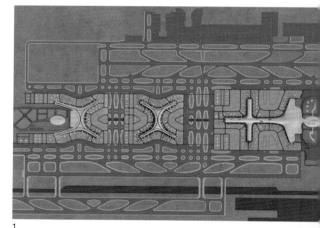

1

2

1   总平面图
    Site plan

2   卫星厅与T3航站楼遥相辉映，共同担当深圳国际门户枢纽
    Satellite concourse and T3 echo to each other, contributing to making Shenzhen a hub and gateway to the world

3   优化X构型，减少占地，提高机坪效率
    X-shaped configuration is optimized to reduce the footprint and enhance the apron efficiency

4   主楼层与捷运车站以中空连接，空间一目了然
    Void connection between main floor and people mover station is a visually unobstructed space

5   剖面图
    Section

深圳机场新航站区地面交通中心（GTC）
Ground Transportation Center (GTC) of New Terminal Area, Shenzhen Airport

随着航空业的快速发展，深圳宝安国际机场需要新建T3航站楼以满足机场吞吐量的增长。新T3航站楼承担起了空侧的交通运输，同时需要一座地面交通中心来解决陆侧复杂的交通运输。机场交通的特殊性、复杂性及多样性决定了作为主要交通枢纽的交通中心有着重要的地位。快速高效，以人为本，人员流程的清晰性和使用的便捷性是设计的宗旨。

项目为新T3航站楼的交通枢纽，为世界各地到港旅客提供各种交通和附加服务，是一个连接T3航站楼与航空城、轨道交通的一个多元化交通核心，承担着连接地上、地面和地下各类交通设施的任务。设计充分利用周边交通组织，将各类交通工具分流至交通中心各个部位，避免人流混杂。内部人流组织采用人车分流+平层换乘的方式，功能清晰，大大简化了复杂的交通情况，并且合理利用人流为建筑增加商业和餐饮附加值。交通中心同时也是一个"中心广场"，所有的商业和服务设施及各种交通模式都被集中在同一个屋檐的下面，相互交融构成了一个复杂的综合体。

GTC地面交通中心与T3航站楼是航空城的核心，两者设计浑然一体，GTC与T3航站楼的自由形态互相呼应，创造出独特的机场景观，以飞扬的形态向人们展示。GTC平面呈椭圆形，三维立体的双曲金属屋面、外倾斜的框式玻璃幕墙，建筑造型光滑、完整，形态沉稳、大气，充分体现"中心"地位。同时，设计充分利用自然元素，令建筑与环境和谐发展，通过天窗、玻璃幕墙，引入自然光线，充分利用周围自然景观，使建筑充满活力。

With the rapid development of aviation industry, it became imperative for Shenzhen Bao'an International Airport to build a new terminal to cope with the growing passenger traffic. The new Terminal 3 addresses traffic and transportation in the airside; meanwhile, a ground transportation center (GTC) is also required for the sophisticated traffic and transportation in the landside. As the major transportation hub, the GTC is of paramount importance in view of the particular, intricate and mixed airport traffic conditions. The design aims to realize fast, efficient and human-oriented operation, clear pedestrian circulation and convenient use.

The GTC is the transportation hub of the new Terminal 3, providing various transportation and additional services to passengers from all over the world. Connecting Terminal 3 with aviation city and metro station, this multi-dimensional transportation center offers access to transportation facilities above, on and below the ground. The design makes the best use of surrounding traffic organization to properly distribute different transportation modes in the GTC, thus avoid mixed pedestrian circulations. As for internal pedestrian organization, the pattern of "separated pedestrian/vehicle circulations + same level interchange" is adopted to realize clear function layout and simplified transportation conditions and to guide more pedestrians to commercial and F&B facilities. The GTC also functions as a "central plaza", where all commercial and service facilities and transportation modes are centralized and interwoven under the same roof to create a mix-used complex.

GTC and Terminal 3 constitute the core of Aviation City, which are perfectly integrated as a whole in design. Echoing the free building form of Terminal 3, the rising GTC helps create unique airport landscape. Featured by elliptical plan, 3D hyperbolic metallic roof and outwards inclined framed glass curtain wall, the building looks smooth, complete, solemn and generous and fully demonstrates the significance of GTC as a "center". On the other hand, with proper use of the natural elements, the building is in harmony with environment and vitalized by the daylight introduced from skylight and glass curtain and surrounding natural landscape.

1

项目地点：深圳 宝安
设计时间：2009-2010
建设时间：2010-2012
建筑面积：58000m²
建筑层数：1-3层
建筑高度：27m
合作单位：意大利FUKSAS
曾获奖项：2015年度全国优秀工程勘察设计行业奖 二等奖
2015年度广东省优秀工程勘察设计奖 二等奖

Location: Bao'an, Shenzhen
Design: 2009-2010
Construction: 2010-2012
GFA: 58,000m²
Floors: 1-3
Height: 27m
Partner: FUKSAS
Awards: The Second Prize of National Excellent Engineering Exploration and Design Award, 2015
The Second Prize of Excellent Engineering Exploration and Design Award of Guangdong Province, 2015

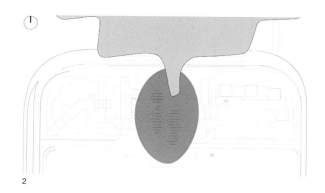

2

1. T3航站楼屋面伸臂与GTC屋面搭接
   Roof of Terminal 3 extends to overlaps with that of GTC

2. 总平面图
   Site plan

3. 二层至三层的中庭
   Atrium from F2 to F3

4. GTC二层换乘大厅,是GTC内部人流"平层换乘"的核心空间,白色的设计基调,使空间简洁、明亮、协调;六边蜂窝形顶棚形态独特而壮观;虚实变化的顶棚与条形采光窗紧密结合,创造出光影变幻的效果
   GTC's transfer concourse on F2 is the core space of "same-level transfer" for internal pedestrian, where the prevailing white tone helps create a concise, bright and well coordinated space with unique and amazing hexagonal honeycomb ceiling; the illusionary change of ceiling and strip skylights jointly create the play of light and shadow

深圳蛇口邮轮中心
Shekou Cruise Center, Shenzhen

深圳市南山区蛇口自贸区太子湾片区，项目作为深圳连通香港、走向世界的"海上门户"，建成集国际邮轮母港、港澳及国内高速客轮港于一体的口岸港务交通航站楼，是亚洲最大型、最新型的邮轮中心综合体，提供海陆交通、口岸联检、服务、办公、商业配套于一体。

建筑整体形态自然呈三角形，充分体现了建筑对城市、对地块的尊重：位于海侧的直角边服务于邮轮母港，另一边服务于高速客轮港，位于陆侧的斜边为中央大厅的主入口。建筑设计根据城市综合体理念，注重空间组织和动线设计。内部空间组织清晰：中央人流动向，两翼的空间既独立又可通过空中连廊便捷地联系起来；中央阶梯贯通公共区楼层；室内平台层层递退，同层空间的水平延展与不同层的纵向贯通灵活地调节各种功能的分配及空间的舒适感受。

建筑形态设计灵感来源于"船首波"，船行航迹和激起浪花的形态得到巧妙的象征和体现，呈现出动感波浪形状的建筑造型，既和周边海域的景观融为一体，又形成了独特的艺术风格，同时蕴含了深圳改革开放乘风破浪、披荆斩棘的创业精神。立面创意上引入海洋"珊瑚"的元素，五彩缤纷的海洋珊瑚的无规则而自然的肌理元素形态呈现给人一种自然优美的形态，使建筑形态极具识别性以及独特性，同时使建筑与海洋的有机结合得到进一步的体现，同时也给人带来建筑驶入海洋的联想。室内空间形态、细节设计、标识设计均延续了海洋元素的应用，参数化控制的曲线层叠形态呈现出轻快的"轻浪、浅滩"的意境。

Located in the Prince Bay of Shekou Free Trade Area, Nanshan District, Shenzhen, the project is envisioned as the city's maritime gateway to Hong Kong and the outside world. It is a port administration and transportation terminal that serves both the international cruise home-port and the port of high-speed passenger ships to and from Hong Kong, Macao and other Chinese cities. As the largest and latest cruise complex in Asia, it integrates the functions of sea and land transport, port joint inspection, service, office and retails.

The naturally shaped triangle building shows full respect for the urban context and site. The sea-side leg serves cruise home-port with the other leg for the high-speed passenger ship port, while main entrance of concourse is provided on the land-side hypotenuse. Following the concept for urban complex, the architectural design focuses on the space and circulation organization. The interior space is clearly organized. The two wings are independent yet conveniently connected via sky bridge. The central stairs run through floors in public area; the interior terrace sets back by floor. The horizontal stretch of the same-floor space and vertical connectivity of different floors flexibly adjust the functional distribution and comfortable spatial experience.

Inspired by the bow wave, the waving building shape ingeniously represents the ship track and surged waves. It well integrates into the surrounding seaside landscape with unique artistic style, while communicating the spirit of braving wind and waves in Shenzhen's reform and opening-up. The irregular and spontaneous texture of colorful coral incorporated in the façade looks natural and attractive, making the building highly recognizable and distinctive and integrate with the ocean in a more organic way, as if the building sets sail to the sea. The interior spatial form, details and signage also continue the marine style with parameterized tiered curves to create the atmosphere of "soft wave and shoal".

项目地点：深圳市前海蛇口自贸区太子湾片区
设计时间：2013-2014
建设时间：2014-2016
建筑面积：138169m²
建筑层数：10层
建筑高度：60m
合作单位：法国岚明建筑设计事务所
　　　　　（Agence Denis Laming）

Location:	Prince Bay, Qianhai Shekou Free Trade Area, Shenzhen
Design:	2013-2014
Construction:	2014-2016
GFA:	138,169m²
Floors:	10
Height:	60m
Partner:	Agence Denis Laming (France)

1

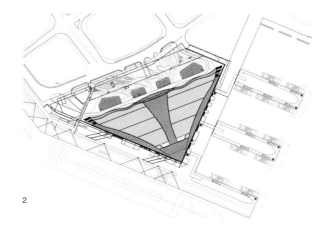

2

1 "海珊瑚"—海洋珊瑚的无规则而自然的肌理元素再现于建筑表皮
 "Sea coral" – building skin recreating the fractal and natural fabric of sea coral

2 总平面图
 Master plan

3 "船首波"—船行的航迹和动感波浪形状的建筑造型
 "Bow Wave" – building shape representing the ship track and dynamic waves

4 现代化的码头,静待邮轮的归港
 Modern terminal welcoming homing cruise

5/6 立面图
 Elevation

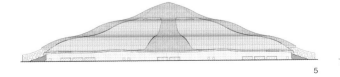

7

8

9

7 舒适大气的联检大厅
　Comfortable and spacious hall of joint inspection

8 明晰易达的联检入口
　Distinctive and easily accessible entrance of joint inspection

9 层层递退的处理方式，同层空间的水平延展与不同层的纵向贯通，灵活地调节各种功能的分配及空间的舒适感受
　With the setback by floor, the horizontal stretch of same-floor space and vertical connectivity of different floors flexibly adjust the functional distribution and offer comfortable spatial experience

10 简洁明快的办票空间
　　Concise and lively check-in space

11 邮轮元素的观海咖啡厅旋转楼梯
　　Cruise-style sea-view café and spiral stair

12 生动活力的登船廊桥
　　Dynamic boarding bridge

武汉火车站
Wuhan Railway Station

作为武汉市的门户建筑,武汉站犹如一只展翅的大鸟,寓意千年鹤归、九省通衢及中部崛起。其"大鹏展翅"的形象不仅映衬着"白云黄鹤"的历史文脉,更以展臂欢迎的姿态表达着城市的态度。武汉火车站于2009年12月26日建成启用。

武汉站为武广客运专线湖北段的客运枢纽站及3个始发站之一(另外2个为新长沙及新广州)。客运站房规模庞大,建筑面积超过10万平方米,2020年每年旅客发送量将达到1750万人次,每日办理旅客列车162对,2030年旅客年发送量可达3100万人,高峰小时旅客发送量9300人。站内设有站台11座,共有20条股道,包括4条正线和16条到发线,是集高铁、公路、地铁、出租汽车、社会车辆于一身的交通建筑综合体,高效组织多种客流流线。武汉站首创等候式和通过式相结合的进站流线模式,"高架候车,上进下出",实现了各种交通工具"零距离换乘",形成铁路与城市公交、地铁的无缝连接,给乘客提供最大方便。该火车站还是规划中的城市轨道交通4号线、5号线的终点站。乘客不出站即可转乘地铁。以客运流线为主干,武汉站设计出丰富的内部空间,围绕巨大的中庭,两侧分布各区候车厅,室内良好的空间尺度感和光环境使得建筑本身舒适而节能。车站主体充分利用拱形结构的稳定性,并进行造型的艺术处理和构件的标准化设计,实现稳定的纵横双向连续拱跨体系,衍伸出"树状"钢结构和混凝土结构造型的核心元素。

武汉站不仅融入了丰富的文化内涵,更在技术和艺术上达到了统一。2012年,运营3年的武汉站获得芝加哥雅典娜建筑设计与博物馆颁发的"2012年国际建筑奖",成为"世界最新最美的建筑"。

项目地点:湖北省武汉市
设计时间:2003-2005
建设时间:2007-2009
建筑面积:客运用房106841m²
建筑层数:地下2层,地上3层
建筑高度:58m
合作单位:中铁第四勘察设计研究院集团有限公司(总包设计单位)
　　　　　AREP公司(外方设计单位)
曾获奖项:2010年度铁路优质工程勘察设计一等奖
　　　　　2011年评为中国"百年百项杰出土木工程"
　　　　　2011年中国土木工程詹天佑奖
　　　　　芝加哥雅典娜建筑设计博物馆颁发"2012年国际建筑奖"

As a gateway to the city, Wuhan Railway Station appears like a gigantic wing-spreading bird which symbolizes the return of the legendary yellow crane, celebrates the city's prominent position as a transportation hub and responds to the national strategy of Rise of Central China. It echoes to the legend about the white cloud and yellow crane on one hand, and presents a welcoming gesture to the visitors on the other hand. The Station was completed and put into use on December 26, 2009.

The Station is the passenger transportation hub of Wuhan-Guangzhou High-Speed Railway at Hubei Section and one of its three departure stations (the other two being New Changsha Station and New Guangzhou Station). With a GFA of over 100,000 m², the Station will handle a passenger traffic up to 17.5 million by 2020 with 162 pairs of in-service passenger trains per day, and 31 million by 2030 with 9,300 passengers per peak hour. The Station contains 11 platforms and a total of 20 tracks including 4 main and 16 arrival-departure tracks. As a transportation complex integrating the traffic of high-speed railway, highway, subway, taxi and private cars, the Station can efficiently organize various passenger circulations on different levels. It is the first to employ a "waiting + passing" entry model, where passengers can wait for the train on the elevated platform and exit at the lower level. The Station offers "zero distance interchange" for seamless connection with bus and subway. As the Station also serves as the terminal of the planned metro Line 4 and 5, the railway passengers can conveniently take the subway right within the station. Focusing on passenger circulation, diversified interior spaces are created to center around the huge atrium flanked by waiting halls on both sides. Appropriate scale and lighting ambient of these interior spaces help create pleasant and energy-efficient built environment. The station building employs stable arch structure which, coupled with artistic form design and standardized components, helps realize a sound arch span system that continuously run longitudinally and transversally, and derive such core elements as tree-like steel structure and concrete structural elements.

Wuhan Railway Station incorporates rich cultural connotations, but also achieved unity in architectural art and construction techniques. In 2012, three years after its operation, the Station was awarded The International Architecture Award for 2012 by Museum of Architecture and Design, Chicago Athenaeum, being acclaimed as the newest and most beautiful building in the world.

Location: Wuhan City, Hubei Province
Design: 2003-2005
Construction: 2007-2009
GFA: 106,841m² (passenger transportation facility)
Floors: 2 underground, 3 aboveground
Height: 58m
Partner: China Railway Siyuan Survey and Design Group Co., Ltd. (General Design Contractor)
　　　　AREP (international firm)
Awards: The First Prize for Excellent Railway Engineering Exploration and Design, 2010
　　　　"100 Outstanding Civil Engineering Projects from 1900 to 2010", 2011;
　　　　Tien-Yow Jeme Civil Engineering Prize, 2011;
　　　　The International Architecture Award, 2012 by Museum of Architecture and Design, Chicago Athenaeum

1　总平面图
　　Site plan

2　夜色下的武汉站,与湖面倒影相映生辉
　　Wuhan Railway Station and its reflection on the lake at night

3　透过吸音管帘顶棚可以看到一体化设计的主体结构、照明、吊顶系统。
　　Integrated main structure, lighting and ceiling system through acoustic pipe curtain ceiling

4　从剖面图可以看出建筑与结构的极度契合;站厅大空间、站台灰空间有序排列展开
　　Section showing perfect architecture-structure integration; orderly display of generous space in concourse and grey space of platform

5　剖透视展现"通过式"的候车方式
　　Sectional perspective showing through-type waiting style

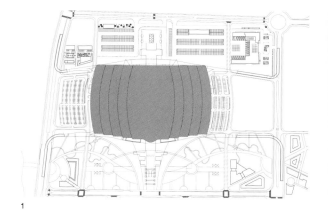

1

6 出发大厅：管帘吊顶有效调节车站内光线，节约运营成本
  Departure hall: Pipe curtain ceiling effectively regulates the light in the station thus cuts operating costs

7 从出发厅可以俯瞰站台
  Overlooking the platform from the departure hall

8 武汉站出发大厅
  Departure hall of Wuhan Railway Station

9 从上至下进站区、站台区、出站区、地铁区层次清晰，使用高效
  Clearly-defined top-down hierarchy, i.e. entrance, platform, exit and metro areas contributing to high efficiency

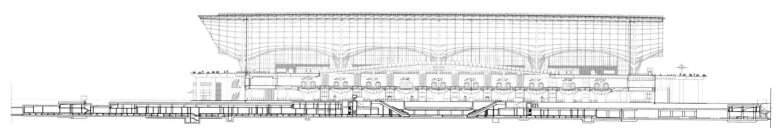

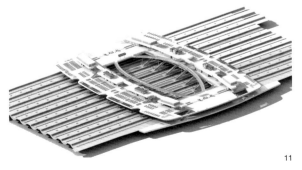

10 旅客候车空间：高侧窗及百叶吊顶引入了足够的自然光线
Passenger waiting space: high-level side window and shutter ceiling ensure sufficient daylight

11 武汉站内部功能布局模型
Internal functional layout model of Wuhan Station

12 吊顶系统及屋面体系与钢结构拱支撑融为一体
Ceiling and roofing system integrating with steel structural arch support

新建云桂铁路引入昆明枢纽工程 新建昆明南站
Kunming South Railway Station (New)

新建昆明南站地处昆明市呈贡区，是集国有铁路、地铁、公交、出租等市政交通设施为一体的特大型综合交通枢纽，是新建云桂铁路引入昆明的枢纽工程。外型设计以"雀舞春城，美丽绽放"的形象寓意象征春城昆明的开放进取和热情好客的城市特质。建筑细部的扇形形体、坡顶木构、浮雕纹饰，都具有浓郁的地域特点和民族特色，象征着多民族的和谐交融及作为西南地区国际友好合作交流的平台。

新建昆明南站总建筑面积33.4万平方米，设计年发送旅客4693万人，为特大型站房。车站共四层，地上三层，地下一层。地上三层是高架层，为主要的旅客候车区域；二层主要为站台层；一层为交换大厅、出站通道和出租车待客区、设备用房；地下一层为地铁1、4号线站台层。昆明南站分为渝昆场、沪昆场、云桂场三个站场共16个站台30条到发线。流线设计采用两侧进站，将站区宽度变为优势，增加了车辆停靠面，大大缩短了旅客步行距离，出站口在地下，可以就近换乘其他交通和地铁。

新建昆明南站利用雄伟的空间巧妙结合高效的通风采光设计，结合昆明温和气候特点，在大空间设计上充分利用热压效应与百叶通风设计，实现了主要的空间均无需机械空调也能保证旅客的舒适性要求。运用候车厅大面积可调节的采光天窗，大大降低了室内照明的运营能耗。建筑外墙面采用白色主体，局部点缀金色，白色是热带亚热带地区建筑的主要色彩，在周围丰富多彩的环境映衬下凸显昆明南站的建筑特点。入夜，当周围环境成为深邃的背景，站房的照明灯光璀璨晶莹、交相辉映，与白天相比又是另一番"七彩云南"的美丽景致。外墙材料采用哑光石材、浮雕石材、隔热玻璃为主，结合钢木结构的入口，防止西向热辐射避免光污染。

The new Kunming South Railway Station, located in Chenggong District, Kunming, is planned as a mega comprehensive transportation hub integrating state-owned railway, metro, bus and taxi systems, as well as the hub that introduces the newly built Yunnan-Guangxi Railway into the city of Kunming. It is designed with the appearance of "a peacock spreading its tails and dancing in the Spring City", which signifies the openness, progressiveness and hospitality of Kunming known as the Spring City to all. Its architectural details, such as fan shape, wooden slope top and embossment are typical of Yunnan characteristics both geographically and ethnically. It serves as a symbol of harmony between various ethnic groups and a platform for genuine international collaboration in the southwest region.

Planned as a mega station building with a total floor area of 334,000m², the Project is designed to handle 46.93 million passengers per year. It has four floors, including 3 aboveground floors (among which F3 serves as an elevated main passenger waiting area, F2 as platform level and F1 as interchange hall, exit passage as well as taxi waiting area and MEP facilities) and 1 underground floor (the platform level of Metro Line 1 and 4). Kunming South Railway Station has three station yards: Yukun Yard, Hukun Yard, and Yungui Yard, totaling 16 platforms and 30 arrival/departure lines. The design allows passengers to enter the station from both sides, making full use of the generous station width. The increased vehicular parking frontage allows the passengers to enter the station without detour. The exit is provided underground to facilitate interchange with other means of transportation including metro.

The new Kunming South Railway Station employs efficient ventilation and daylighting design in its magnificent space. In view of Kunming's mild climate, the design makes full use of thermal pressure and lamella ventilation to ensure comfortable stay of passengers in major spaces even without mechanical air-conditioning. Extensive use of adjustable skylights in the waiting hall greatly reduces energy consumption of interior lighting. The external wall of the building is dominated by white, which is typical of buildings in tropical and sub-tropical zones, and interspersed with gold in some positions. This, coupled with the diverse surrounding environment, highlights the architectural characteristics of Kunming South Railway Station. As evening draws on, the surroundings turn into a deep backdrop while the station building is beautifully lit, revealing another scene of "colorful Yunnan" that is totally different from the daytime one. The external wall is composed of matte stone, embossment stone and insulating glass, which, together with steel-wood entrance, can effectively prevent heat radiation from the west and light pollution.

项目地点：云南省昆明市
设计时间：2010.4
建设时间：2012-2016
建筑面积：334000m²
建筑层数：地上3层，地下2层
建筑高度：52.15m
合作单位：中铁第四勘察设计研究院集团有限公司

Location: Kunming City, Yunnan Province
Design: 2010.4
Construction: 2012-2016
GFA: 334,000m²
Floors: 3 aboveground, 2 underground
Height: 52.15m
Partner: China Railway Siyuan Survey and Design Group Co., Ltd.

1 总平面图
  Site plan

2 外形设计以"雀舞春城，美丽绽放"的形象寓意象征春城昆明的开放进取和热情好客的城市特质
  The appearance of "a peacock spreading its tails and dancing in spring city" signifies the openness, progressiveness and hospitality of Kunming

3 入夜，当周围环境成为深邃的背景，站房的照明灯光璀璨晶莹、交相辉映，与白天相比又是另一番"七彩云南"的美丽景致
  As evening draws on, the surroundings turn into a deep backdrop while the station building is beautifully lit, revealing another "colorful Yunnan" that is totally different from the daytime one

4 鸟瞰昆明南站的雄伟气势
  A Bird's eye view of Kunming South Railway Station

5 鸟瞰车站的竖向关系
  A Bird's eye view of station's vertical relation

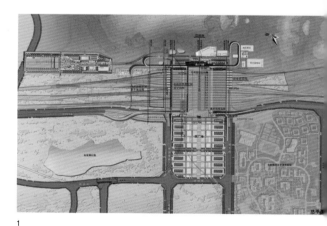

1

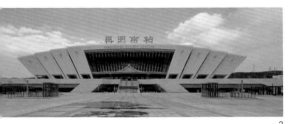

2

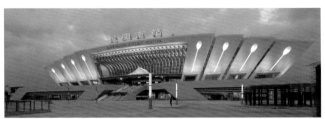

3

4

5

6

7

8

6 具有浓厚民族色彩进站通道
  Entrance passage with strong ethnical colors

7 利用大空间巧妙结合高效的通风采光设计,实现了主要的空间均无需机械空调也能保证旅客的舒适性要求。综合运用候车厅大面积可调节的天窗,采光铺地,以及直通站台的玻璃进站体辅助"引光"降低了室内照明的运营能耗
  Big spaces plus efficient ventilation and lighting design ensure comfortable stay of passengers in main spaces even without mechanical air-conditioning. Extensive adjustable skylight and daylighting pavement in the waiting hall and glass entrance leading to the station platform are used to introduce daylight and reduce energy consumption of interior lighting

8 出站厅以浅色为主色调,简洁、优雅
  Simple and elegant exit hall, dominated by light color

广州市轨道交通四号线车陂南-黄阁段工程
Metro Stations of Guangzhou Metro Line 4

广州市轨道交通四号线是广州第一条以高架线路为主的地铁线路，线路全长69.67公里，其中高架部分48.26公里。

2004年，通过设计招标我院获得了二期工程中的石碁、海傍、低涌、东涌、黄阁汽车城等五座高架车站的设计权。

线路经过亚运城、居住新区、南沙汽车城等不同性质的城市功能组团，承担了区域交通核心的功能。车站建筑对此给予了表达，形成功能模式统一，建筑形态既具有线路共性特征、又具有地域个性特征的可识别性，即广义的"一线一景"建筑形式也与本地气候特征相适应，解决了通风、散热、挡雨、遮阳、采光等基本功能要求，创造出独特的、具有南方特色的高架车站建筑形式。

The 69.67km long Guangzhou Metro Line 4 is the first metro line in Guangzhou reigned by elevated rail of 48.26km.

In 2004, after winning the bid, we were awarded with the design contract for five elevated metro stations, i.e. Shiqi, Haibang, Dichong, Dongchong, and Huangge Auto Town.

Linking up different urban functional clusters ranging from Asian Games Town, new residential area to Nansha Auto Town, Metro Line 4 serves as the major regional transportation means. The station buildings are provided with consistent functional pattern, while building appearances share some common features of this metro line (i.e. the One Typical View for One Metro Line) without sacrificing the characteristics of the locality. The building form of the elevated station is adapt to the local climate, responds to the basic functional requirements of ventilation, thermal discharge, weather protection and daylighting, and presents distinctive local features of South China.

项目地点：广州番禺、南沙
设计时间：2004.3-2005.10
建设时间：2004.8-2006.12
建筑面积：共32587m²
建筑层数：2-3层
建筑高度：18.6-23.3m
曾获奖项：广东省注册建筑师协会第五次(2009年度)优秀建筑佳作奖
2009年度广东省优秀工程勘察设计工程设计三等奖
2009年度广东省优秀工程勘察设计建筑结构专项二等奖
2009年度广东省优秀工程勘察设计建筑环境与设备专项三等奖

Location: Panyu/Huangsha, Guangzhou
Design: 2004.3-2005.10
Construction: 2004.8-2006.12
GFA: 32,587m²
Floors: 2-3
Height: 18.6-23.3m
Awards: The Fifth Excellent Architecture Creation Award by Guangdong Chapter of Association of Chinese Registered Architects, 2009
The Third Prize of Engineering Design under Excellent Engineering Exploration and Design Award of Guangdong Province, 2009
The Second Prize of Building Structure under Excellent Engineering Exploration and Design Award of Guangdong Province, 2009
The Third Prize of Built Environment and Services under Excellent Engineering Exploration and Design Award of Guangdong Province, 2009

1

2

1 位于路侧的石碁站，可以明确看到全线高架车站弧形组合这一共性特征，高低交错的造型也是遵从通风散热功能的实际需求
The roadside Shiqi Station is designed with curved structural combination as the common features of all elevated stations along Metro Line 4. The jagged shape also caters for the actual demand of ventilation and heat dissipation

2 位于亚运城的海傍站，是2010广州亚运会亚运村的交通枢纽，舒缓伸展的屋面为换乘客流提供了遮荫的室外空间
Haibang Station in the Asian Games Town functioned as the transportation hub for the Asian Games Village in 2010. The smoothly extending roof offers shaded outdoor space for transfer passengers

3 高架于道路中的黄阁汽车城站，外置站厅级设备管理用房，最大程度减弱了车站对城市道路景观形成的压迫感
Elevated in the middle of the road, Huangge Auto Town Station with separated concourse-level equipment management facilities minimizes the oppression of the station on the urban road landscape

4 暮色中的低涌站，设备管理用房首层部分架空，显露出车站的南侧入口，Y形柱在减少柱子的同时也增加了建筑的趣味
Dichong Station in dusk twilight, where the ground floor of equipment management rooms are partially opened up to reveal the south entrance; Y-shaped columns reduce the columns while adding interest to the building

Transportation Buildings 交通建筑

广州市轨道交通五号线首期工程-坦尾站
Tanwei Station of Guangzhou Metro Line 5

坦尾站位于广州市荔湾区大坦沙岛中部,北临60米规划双桥道路、南临广佛放射线及广三铁路,站中范围2双桥中路穿越。车站周边交通繁忙,道路、桥梁多,管线极其复杂。5号线东西走向高架线路敷设,紧邻广佛放射线高架路,6号线地下线路南北穿行敷设,5、6号线在此交汇,设置换乘车站坦尾站,同时设置5、6号线联络线。五号线采用高架一层箱梁结构,岛式站台,六号线车站采用明挖一层局部二层框架结构,侧式站台,车站及区间均采用明挖。

车站为广州市轨道交通五、六号线换乘站,是广州地铁线网中首个高架-地下换乘车站;其中五号线为高架线东西走向,六号线则为地下线南北穿行。

车站总建筑面积约为(未含联络线及区间)14217.872平方米,其中地下车站主体建筑面积约为7364.49平方米,高架车站主体建筑面积约为6136.28平方米,附属建筑面积约为717.102平方米。

Located in the middle of Datansha Island, Liwan District, Guangzhou, Tanwei Station adjoins the proposed 60m-wide double-bridge road on the north, Guangzhou-Foshan Lines and Guangzhou-Sanshui Railway on the south, with Shuang Qiao Zhong Lu passing through the station site. The station is encompassed by numerous roads, bridges and pipelines with heavy external traffic. For the convergence of the east-west elevated rail of Line 5 that neighbors elevated road of Guangzhou-Foshan Lines and north-south underground rail of Line 6, Tanwei Station is designed as interchange station with rail link for Line 5 and 6. Line 5 adopts elevated box beam structure and island platform, while Line 6 is characterized by frame structure with one open-cut floor (partially two floors), side platform and open excavation for station and section.

As the first interchange station between elevated and underground rail in Guangzhou metro network, Tainwan Station allows for the interchange between east-west elevated Line 5 and north-south underground Line 6.

The station is designed with a GFA of 14,217.872m² (excluding rail link and section) including 7,364.49m² for underground station, 6,136.28m² for elevated station and 717.102m² for auxiliary structures.

项目地点:广州市荔湾区大坦沙岛中部双桥路
设计时间:2004
建设时间:2004.5-2013.12
建筑面积:14217.872m²
建筑层数:2层
建筑高度:21.15m
曾获奖项:广东省优秀工程勘察设计三等奖

Location: Shuang Qiao Lu in the middle of Datansha Island, Liwan District, Guangzhou
Design: 2004
Construction: 2004.5-2013.12
GFA: 14,217.872m²
Floors: 2
Height: 21.15m
Award: The Third Prize of Excellent Engineering Exploration and Design Award of Guangdong Province

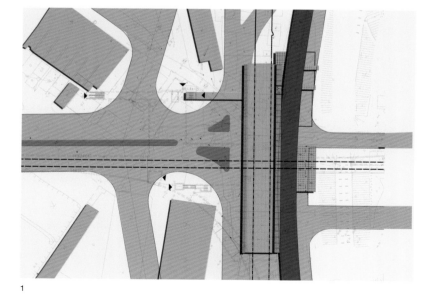
1

2

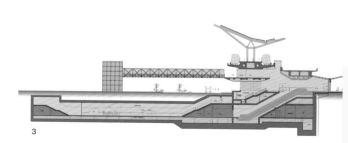
3

1	总平面图	Site plan
2	坦尾站全景	Full view
3	剖面图	Section
4/5	坦尾站外景	Exterior view
6/7	车站站内实景	Interior view

Transportation Buildings 交通建筑

广州市轨道交通五号线工程-动物园站
Zoo Station of Guangzhou Metro Line 5

动物园站位于广州市老城区中较为繁华、交通繁忙的城市主干道环市东路与梅东路交界路口，广州动物园南门前广场地下，车站建筑充分考虑对周边环境的影响，因地制宜，受场地限制及"S"形区间线路转弯等因素影响，车站方案充分利用广场空地布置明挖站房，在环市东路地下暗挖双层叠线侧式站台，通过横通道连接站房、站台。车站建成有效缓解周边商住区交通压力，方便游客游玩动物园。

动物园原生态的自然环境及地域特色，激发了车站设计以清水混凝土作为主题的灵感，清水混凝土粗犷、原始的材质特质呼应车站所处地域环境，采用大红色搪瓷钢板为主色配以精心的室内照明设计，进一步突出了动感活力的地域元素。

车站建筑利用深埋三层明挖站厅形成中庭式空间，为乘客提供视线开敞的交通导向，双"丫"柱结构，成为高大站厅空间的视觉焦点，最大限度地改善地下封闭空间的沉闷和压抑感。车站空间宽敞、简洁、明快、朴实、经济，体现现代交通建筑的特点。

Located at the intersection of two urban main roads, Huan Shi Dong Lu and Mei Dong Lu in the bustling downtown with heavy traffic, the station is underneath the square in front of the south entrance of Guangzhou Zoo. The project design gives due consideration to the impact on the surroundings, the site constraints and the turning of s-shaped rail alignment. The open space of the square is fully used for construction of the open-cut station building, while the double-lapped side platform is excavated underneath Huan Shi Dong Lu. The station building and platform are then connected via the traverse passage. The completed station effectively alleviates the traffic pressure of surrounding commercial and residential area and facilitates the visit to the Zoo.

Inspired by the primitive natural environment and regional features of the Zoo, thematic element of fair-faced concrete with rough and pristine texture in turn responds to the local environment of the station. The prevailing enameled pressed steels in red, along with the carefully designed interior lighting, further highlight the dynamic and vitality.

The atrium-like space created in the open-cut concourse with the burial depth of three floors offers visually unobstructed orientation to the passengers. The eye-catching double Y-shaped columns in the lofty concourse minimize the depression brought by enclosed underground space. The spacious, concise, bright, pristine and cost effective station space sets up an example of modern transportation building.

项目地点：广州市环市东路动物园南门
设计时间：2004.7-2009.12
建设时间：2009.12
建筑面积：17906.36m²
建筑层数：地下4层
建筑高度：30.9m(地下)
曾获奖项：广东省优秀工程勘察设计三等奖

Location: South gate of the Zoo, Huan Shi Dong Lu, Guangzhou
Design: 2004.7-2009.12
Construction: 2009.12
GFA: 17,906.36m²
Floors: 4 underground
Height: -30.9m
Award: The Third Prize of Excellent Engineering Exploration and Design Award of Guangdong Province

1   总平面图
    Site plan

2   典型剖面关系图
    Typical sectional relation

3   站台空间，暗挖单洞双层叠线侧式站台，通过横通道连接站厅
    Platform space of the subsurface-excavated double-lapped side platform with single cavern connecting to the concourse via the traverse passage

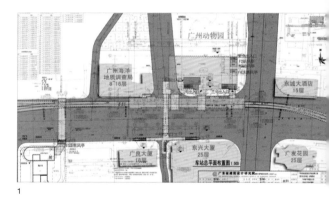

1

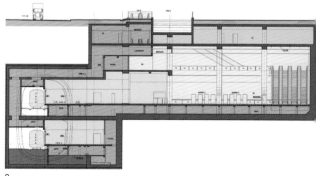

2

3

4

5

4   站厅中庭顶棚节点,动物园原生态的自然环境及地域特色,激发了设计以清水混凝土作为主题的灵感,清水混凝土粗犷、原始的材质特质呼应车站所处地域环境,站厅中庭以室内背光照明设计,进一步突出了动感活力的地域元素
Concourse ceiling and atrium node; inspired by the primitive natural environment and regional features of the zoo, the thematic element of fair-faced concrete with rough and pristine texture responds to the local environment of the station. The backlighting of the atrium further highlights the dynamic and vitality

5   站厅中庭空间,利用深埋三层明挖站厅形成中庭式空间,为乘客提供视线开敞的交通导向,配以双"Y"柱结构,成为高大站厅空间的视觉焦点,改善地下封闭空间的沉闷和压抑感
Atrium space of concourse; the atrium created in the open-cut concourse with the burial depth of three floors offers visually unobstructed orientation for the passengers. The eye-catching double Y-shaped columns in the lofty concourse minimize the depression brought by enclosed underground space

体育建筑

随着中国的经济发展，体育运动逐渐普及，中国先后举办了具有国际影响的奥运会、亚运会、各种国际锦标赛等各种盛会，这些体育盛会，对体育建筑的发展有了很好的推动作用。体育场馆，由于其设计独特、体量巨大、具有较强的公共性，往往会成为城市或区域的地标，成为提升城市公共空间品质的重要节点。

GDAD前后参与了奥运会、亚运会等多个场馆的设计，在这些场馆建设中发挥了积极的作用。这些项目设计权的获得，都是通过国际或全国方案设计竞赛的遴选。GDAD的设计团队在这些设计竞赛中展示了优秀的方案设计原创能力，多次中标并实施。

这些项目包括2010年广州亚运会唯一新建的主场馆——广州亚运馆、以表现速度与力量为主题的2008年奥运会老山自行车馆、以亚运会五羊会徽与极限运动头盔形象相融合的广州自行车馆以及作为省运会比赛场地的佛山体育馆、惠州市金山湖游泳跳水馆、肇庆新区体育中心等体育场馆。尤其是广州亚运馆、老山自行车馆、广州自行车馆均为国际竞赛设计独立原创中标实施的优胜作品，这在众多国际设计机构在国内高端项目方案设计上几乎形成垄断的背景下殊为不易。

这些设计作品，涵盖了体育场、体操馆、台球馆、壁球馆、跳水馆、自行车馆等多种综合及专项的体育场馆，涉及的建筑功能及体育工艺均较为复杂、公共性强，具有多种人流动线、人流量大、消防疏散设计要求高；都是大空间建筑，具有大跨度钢结构和复杂多变的屋面墙面围闭体系；由于其建筑的标志性，涉及的建筑造型变化丰富，具有多种建筑材料的组合和精细的建筑节点；同时这些也都需要大空间空调、大空间照明、屋面排水等复杂的机电工艺设计；这些设计的技术和集成程度较高，涉及多部门协同设计，配合过程复杂，需要较高的系统配合及协调能力；这些项目现场施工要求高、工序交接繁多、施工组织复杂，需要很好的现场配合服务和攻关解决施工或材料工艺复杂现场问题的能力；这些场馆的设计都具有较高完成度，最终呈现的较好实施效果，从而为所在城市贡献了一个个具有较高标志性的公共建筑，为城市空间的优化和提升起到积极的作用。这些项目的设计和实施都充分体现了GDAD各专业设计配套齐全、系统垂直整合能力高、服务配合意识较强的优势。

GDAD注重设计创新，在新技术、新材料应用方面做了很多有益探索，并善于结合项目积极推广应用，例如广州亚运馆充分利用计算机三维模拟技术，完成钢结构、金属屋面板、玻璃幕墙及金属幕墙的设计；在材料、结构等方面进行了新的探索，大面积使用隐藏拉索式双曲玻璃幕墙，通过整体分析整合幕墙的构造、节点设计，使幕墙与主体结构完美结合；金属屋面系统为445R铁索体高耐候性不锈钢金属屋面板为国内首次全系统应用；在结构设计中由于积极创新，从而开发出三个专利技术等等。

Sports Buildings

Thanks to the rapid economic growth and increasing popularity of sports in the country, China has hosted a string of important international sports events including Olympic Games, Asian Games and various international championships, greatly driving the development of sports building. Sports venues, with distinctive design, massive scale and strong public nature, can easily become the regional or citywide landmarks and key node to upgrade the quality of urban public space.

GDAD has participated in a number of international or nationwide design competitions for the venues of the Olympic Games and Asian Games, and won the design contracts with outstanding creative design that are finally implemented.

Those projects include Guangzhou Asian Games Gymnasium, the only new main venue for 2010 Guangzhou Asian Games, Laoshan Velodrome themed on speed and strengthen for 2008 Beijing Olympic Games, Guangzhou Velodrome integrating the elements of five-ram emblem of Asian Games and x-sports helmet, as well as the sports venues for Guangdong Provincial Sports Games like Foshan Gymnasium, Huizhou Jinshan Lake Swimming and Diving Natatorium and Zhaoqing New Area Sports Center. For the first three projects, we won the design contracts through international competitions with our own creative design, which was quite unusual considering the fact that international design firms almost monopolized the design of key projects in China.

These design projects cover stadium, gymnasium, billboard center, squash center, natatorium, velodrome, as well as other comprehensive and specialized sports venues. They involve complicated building functions and sports technology, distinct public nature, multiple pedestrian circulations, dense pedestrians and high fire evacuation demands. These large-space buildings usually requires large-span steel structure and complicated roof, wall and envelope system. Due to their iconic features, they also present varied and distinctive building forms which are realized through combination of different materials and delicate building nodes. Besides, they are also supported by large-space AC and lighting, roof drainage and other complicated MEP systems. The high technical and integration levels require effective collaboration and coordination among various departments. The high-demanding construction, cumbersome delivery procedures and complicated construction organization call for efficient construction administration and effective solution to construction, material or technology problems encountered at site. These projects have been implemented to a satisfactory completion level and desirable effect, presenting highly representative public buildings for their cities and contributing to the improvement and upgrading the urban space. The design and implementation of these projects fully showcase our strength in full-fledged disciplines, effective vertical integration and professional service and cooperation.

Attaching importance to design innovation, we have been ceaselessly exploring the applications of new technologies and materials, and promoting the applications through project practices. For example, in the design of Guangzhou Asian Games Gymnasium, we employed the computer 3D simulation technology for the engineering of steel structure, metallic roof, glass curtain wall and metallic curtain wall. We also tried the new material and structure applications. We use extensive hidden cable hyperboloid glass curtain walls to seamlessly integrate the curtain wall with the main building based on overall analysis of the construction and node design of the composite curtain wall. The metallic roof system made of 445R ferrite highly anti-weathering stainless steel plate was firstly applied in China. Eventually three patented technologies were developed out of the various structural innovations.

广州亚运馆
Guangzhou Asian Games Gymnasium

广州亚运馆（原名广州亚运城综合体育馆），是2010年广州亚运会唯一新建的主场馆，本届亚运会标志性建筑。亚运馆凭借其创新设计理念、独特的建筑体验、标志性和可实施性的绝佳平衡，在国际设计竞赛中获胜并且成为实施方案。

亚运馆包括了体操馆、综合馆和亚运博物馆等一系列功能。区别于传统体育场馆的设计方法，设计方案大胆的利用有机连续的金属屋面板统领多个场馆，场馆犹如珍宝隐藏于屋面之下。场馆间、屋檐下的灰空间连贯舒展，体现了传统建筑文化和场所精神，大面积金属屋面轻盈飘逸，独具岭南建筑的神韵。设计方案注重营造积极开放的城市公共空间，设置穿越场馆的城市广场，为城市带来新活力，提高其公共性。整个亚运馆造型生动，利用445R铁素体高耐候性不锈钢金属屋面板系统，解决了复杂的屋面排水问题，完美的实现设计团队期待的屋面曲线，从城市空间各个角度呈现出不断变化的造型以及色彩变化，展示亚运馆丰富的建筑形象。亚运馆充分利用计算机三维模拟技术，完成钢结构、金属屋面板、玻璃、金属幕墙的设计。在材料、结构等方面进行了新的探索：隐藏拉索式双曲玻璃幕墙幕墙二次结构与主体结构完美结合；清水混凝土浇筑面积达25000平方米等，这些探索及成果效果受到各方高度评价。

广州亚运馆设计与施工历时逾两年，跨越3载，对于一个如此复杂的大跨度三维连续曲面的体育建筑，这是巨大的挑战。在设计团队的共同努力下，仍然达到了极高的设计完成度，在新技术、新材料应用做了很多探索，在自主创新设计等方向实现了突破。随着项目竣工验收并投入测试赛，广州亚运馆以其梦幻般的形象和独特的体验迅速获得社会各界的高度关注，赢得赞誉！

Guangzhou Asian Games Gymnasium (formerly Guangzhou Asian Games Town Gymnasium), the iconic building for the Asian Games, is the only main venue newly built for 2010 Guangzhou Asian Games. With innovative design concept, unique architectural experience and equilibrated iconicity and viability, our design proposal won the international design competition and was finally implemented.

The Gymnasium consists of Gymnastic Hall, the Gym, the Asian Games Museum etc. Unlike conventional sports venue, the free-flowing continuous metallic roof is designed to unify the various facilities, while the venues appears to be some hidden treasures covered by the roof. The smoothly extended grey spaces under the eaves link up all facilities, reflecting the traditional architectural culture and spirit of place. The large-sized yet light and elegant metallic roof reveals the style of Lingnan architecture. Emphasizing on creating open urban public space, we provide urban square that runs through the venue to inject new momentum into the city and enhance the public nature of the venue. The dynamic form of the Gymnasium is constructed with the metallic roof system comprising 445R ferritic stainless steel of high weatherability. This material can help tackle the complicated drainage issues, realize the desired roof curve and present a changing building image with varied building forms and color when viewed from different angles of the urban space. 3D simulation technology is fully employed in design of steel structure, the metallic roofing system and the glazed/metallic façade. There are also some explorations on new materials and structures. For instance, the secondary structure of hyperbolic glazed facade supported by hidden stayed cables perfectly combines with the main structure. The fair-faced concrete amounts to 25,000m^2. Those explorations and results are highly appraised by all walks of life.

It is a huge challenge to design and build such a sophisticated sports venue with continuous large-span 3D curvilinear roofing within 3 years. Thanks to the concerted efforts of the design team, the project is completed to the desired effect with numerous explorations on new technologies/ materials and independent innovations. Since its completion and operation for test event, the Gymnasium has been highly acclaimed for its fascinating presence and unique experience.

项目地点：广州 番禺
设计时间：2007-2008
建设时间：2008-2010
建筑面积：65315.0m^2
建筑层数：4层
建筑高度：33.8m
曾获奖项：2008年国际竞赛优胜奖中标
　　　　　AAA2014亚洲建筑师协会奖：专业建筑类别荣誉奖
　　　　　2011年全国优秀工程勘察设计行业一等奖
　　　　　2011年荣获百年百项杰出土木工程
　　　　　2011年度中国建筑金属结构协会颁发的中国钢结构金奖（钢结构工程结构设计优秀奖）
　　　　　2011年评为中国第十届"詹天佑土木工程大奖"
　　　　　2011年詹天佑土木工程奖创新集体奖
　　　　　2011年度第六届"中国建筑学会建筑创作优秀奖"
　　　　　2013年香港建筑师学会两岸四地建筑设计大奖优异奖
　　　　　2011年度广东省优秀工程勘察设计奖一等奖
　　　　　住房和城乡建设部2009年绿色建筑与低能耗建筑"双百"示范工程
　　　　　2010年度China-Designer中国室内设计年度评选年度优秀公共空间设计金堂奖
　　　　　2011年度广东省注册建筑师优秀建筑创作奖
　　　　　2011年度广东省空间结构学会颁发的广东钢结构金奖"粤钢奖"设计奖一等奖

Location: Panyu, Guangzhou
Design: 2007-2008
Construction: 2008-2010
GFA: 65,315.0m^2
Floors: 4
Height: 33.8m
Awards: Winning proposal of international competition in 2008;
The ARCASIA Award (AAA2014): Honor Award – Architecture
The First Prize for National Excellent Engineering Exploration and Design - Engineering Exploration and Design, 2011
"100 Outstanding Civil Engineering Projects from 1900 to 2010", 2011
Gold Prize for Steel Structure (Excellence in Steel Structure Design) by China Construction Metal Structure Association, 2011
The Tenth Tien-Yow Jeme Civil Engineering Prize, 2011
Innovation Collective Award under Tien-Yow Jeme Civil Engineering Prize, 2011
Excellent Award of the 6th ASC Architectural Creation Award, 2011
Excellence Award under Cross-Strait Architectural Design Awards by Hong Kong Institute of Architects, 2013
The First Prize of Excellent Engineering Design Award of Guangdong Province, 2011
Model Project for "100 Top Green Buildings and 100 Top Energy-saving Buildings" by Ministry of Housing and Urban-Rural Development of the PRC, 2009
Jin Tang Prize of Excellent Public Space Design under China Interior Design Awards, 2010
Excellent Architecture Creation Award by Guangdong Chapter of Association of Chinese Registered Architects, 2011
The First Prize of Gold Award for Steel Structure Design in Guangdong (Yuegang Award) by Guangdong Provincial Society for Spatial Structures, 2011

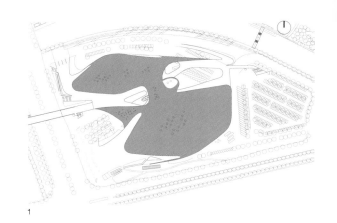

1

1 总平面图
   site plan

2 亚运馆概念设计手稿
   Concept sketch

3/4 有机连续的金属屋面外壳统领各个场馆
    Organically continuous metallic roof shell unifies various facilities

Sports Buildings 体育建筑 175

5

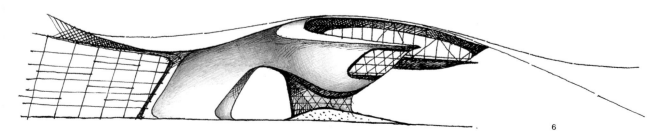

6

176 体育建筑 Sports Buildings

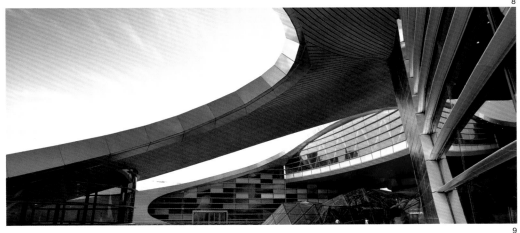

5 受当地传统建筑的影响,亚运馆飘檐出挑大,呼应了该区域的气候,从而降低建筑能耗
Inspired by local traditional architecture, the generously overhung roof echoes with the local climate and cut the building's energy demand

6 亚运馆空间节点设计手稿
Spatial node sketch

7 亚运历史展馆入口采用结晶体驳接爪件点式幕墙,造型独特
The unique façade at the entrance to the Asian Games History Exhibition Hall is of distinct and unique form

8 从二层平台灰空间看亚运历史展馆入口
The entrance to the Asian Games History Exhibition Hall viewed from grey space of F2 platform

9 三维屋面营造了流动连贯的灰空间
Flowing and continuous grey space created by 3D roof

2008年奥运会北京老山自行车馆
Laoshan Velodrome for 2008 Beijing Olympic Games

2008年奥运会老山自行车馆,是国家体育局为举办第29届奥运会而兴建的专业自行车比赛馆,与鸟巢、水立方、北京射击馆同为2008年北京奥运会首批兴建的四个大型场馆之一。该馆提供一个可容纳6000观众同时观看场地自行车比赛的场所,同时能满足残奥会的使用要求。

根据地形特点,老山自行车馆设计采用圆形主体与低平的裙房相结合的平面布局,33米高的碟形建筑主体设于用地南侧,其与西面的五环路之间形成了开阔的休憩广场,便于赛时各种流线的组织,且在五环路上有良好的视觉景观。比赛大厅位于二层,用于内部运营、管理、技术支持用房均集中于一层,减少了与赛场的交叉干扰,对赛后使用及管理也提供了方便条件。

主体造型以表现速度与力量为主题,在2008年奥运会大型场馆的方案投标中率先实现了由中国本土设计师自主设计中标实施。该建筑设计创造性地把巨型圆环桁架梁与人字形柱结合,使大跨度结构悬在空中,为比赛大厅提供了一个优质的巨大空间和良好的自然光线。大跨度结构无侧推支座,优化了看台底部的采光通风效果。同时为节约建筑用材成本便于赛后运营作出了比较突出的贡献。奥运会期间,受到社会的广泛关注与好评。

Laoshan Velodrome for 2008 Beijing Olympic Games is a professional cycling venue developed by the General Administration of Sport of China for the 29th Olympic Games and one of the four large venues firstly built for 2008 Beijing Olympic Games apart from Bird's Nest (the National Stadium), Water Cube (National Aquatics Center) and Beijing Shooting Range Hall. The 6,000-seat Velodrome for track cycling is also available for the Paralympic Games.

Following the topography, Laoshan Velodrome adopts the planar layout that combines the circular main building with the low and flat podium. A spacious square is framed by the 33m high dish-shaped main building in the south and the 5th Ring Road to the west to facilitate the organization of various circulations during the game and provide attractive landscape viewed from the 5th Ring Road. The competition hall is located on F2, while the internal operation, management and technical support rooms are centralized on F1 to minimize the intersection and interference with the competition field and facilitate the post-game operation and management.

Themed on the speed and strength, this design proposal independently developed by local Chinese architect for the first time won the bids for large venues for 2008 Beijing Olympic Games and was eventually implemented. The architectural design creatively combines massive loop truss beam with the herringbone column to realize a suspended large span structure, allowing for a quality large space and favorable daylighting for the competition hall. The large span structure free of lateral support improves the daylighting and ventilation at the bottom of stand and greatly helps to lower the building material costs and facilitate the post-game operation. The Velodrome enjoyed tremendous popularity during the Olympic Games.

2

项目地点: 北京市石景山区
设计时间: 2004
建设时间: 2004-2008
建筑面积: 32500m²
建筑层数: 1层
建筑高度: 33m
合作单位: 中国航天建筑设计研究院
曾获奖项: 2008年度全国优秀工程勘察设计奖铜奖
2008年度全国土木工程詹天佑大奖
2008年度北京奥运会工程优秀勘察设计奖
2005年度广东注册建筑师创作奖

Location: Shijingshan District, Beijing
Design: 2004
Construction: 2004-2008
GFA: 32,500m²
Floors: 1
Height: 33m
Partner: China Aerospace Architectural Design Academy Group
Awards: Bronze Award of National Excellent Engineering Exploration and Design Award, 2008
                Tien-Yow Jeme Civil Engineering Prize, 2008
                Excellent Engineering Exploration and Design Award for Beijing Olympic Games (2008) Projects
                Architecture Creation Award by Guangdong Chapter of Association of Chinese Registered Architects, 2005

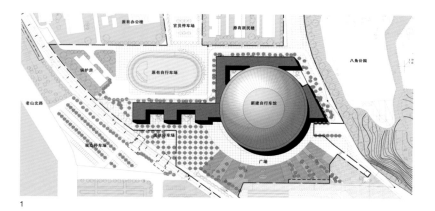

1

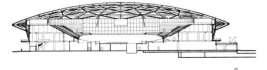

1 总平面图
　Site plan

2 从南广场看自行车馆主体，舒展的体型和严谨的构件展现了建筑美学和结构形式的结合
　Viewed from South Square, the main building with sweeping form and precise components presents perfect integration of architectural aesthetics and structure

3 主体结构大跨度出飘形成了两层高的室内外过渡空间
　Large span overhanging of main building creates two-floor transitional space between interior and exterior

4 自行车馆室外的小空间
　Small space outside Velodrome

5 自行车馆东侧的道路
　Road on the east

6 剖面图
　Section

7 北立面
　North façade

Sports Buildings 体育建筑　179

广州自行车馆
Guangzhou Velodrome

广州自行车馆为2010年广州亚运会场地自行车赛和花样轮滑表演赛比赛场地，是省院独立投标并以国际设计竞赛第一名中标实施的设计方案，为华南地区首座国际标准室内自行车赛馆。

广州自行车馆造型设计概念为亚运会五羊会徽与极限运动头盔形象的融合。屋面造型为透雕效果的椭圆球体，曲线流畅，形态舒展。建筑参考岭南传统建筑的敞厅和骑楼做法，设计了敞开式观众大厅和环馆遮阳通廊，将地方传统建筑的智慧融汇到现代建筑创作中。设计中考虑了多重节能手法，降低后续运营能耗。

自行车馆建筑面积26865平方米，地上三层，固定座席1780座，活动座席240座。其结构形式为预应力钢筋混凝土框架结构，屋面为局部双层钢网壳结构，网壳长度达102米（短轴）- 126米（长轴），是目前国内超限大跨度局部双层单层网壳结构的首例。自行车馆内设宽7.5米，长250米的国际标准自行车木赛道，该赛道设计具有非常高的技术含量，是目前国际最优秀的室内自行车赛道之一，其精度测量误差仅2.4毫米，在2010年亚运比赛中，这条赛道上共产生了6项新亚洲纪录，为刷新纪录最多的比赛项目。

As the first international standard indoor velodrome in South China, Guangzhou Velodrome was planned for the track cycling event and figure roller skating exhibition competition for 2010 Guangzhou Asian Games. This design proposal independently developed by us was finally implemented after winning the first place in the international design competition.

The building shape is envisioned by integrating the elements of five-ram emblem of the Asian Games and x-sports helmet. The roof in the form of engraved ellipsoid features free-flowing curves and form. By referencing the open hall and Qilou (arcade) in traditional Lingnan architecture, the open auditorium and encircling sunshade corridor are designed as a way to inject the wisdom of local traditional architecture into the modern architectural design. Besides, several energy efficiency approaches are applied to cut the energy consumption in post-game operation.

The 3-floor velodrome with a GFA of 26,865m^2 offers 1,780 permanent seats and 240 movable seats and adopts pre-stress reinforced concrete frame structure. The partial double-layer steel latticed shell in the length of 102m (short axis)-126m (long axis) for the roof is the first over-the-code large span single-layer latticed shell with partial double-layer applied in China. The 7.5 wide and 250m long international standard cycling track is highly technical and one of the top indoor cycling tracks worldwide with the error of only 2.4mm in precision measurement. During the 2010 Asian Games, this track witnessed 6 new Asian records, dwarfing other competition events.

项目地点：广东省广州市
设计时间：2008
建设时间：2008.4-2010.10
建筑面积：26856m^2
建筑层数：3层
建筑高度：39.06m
曾获奖项：2011年度全国优秀工程勘察设计行业奖二等奖
2011年度广东省优秀工程勘察设计奖二等奖

Location: Guangzhou, Guangdong Province
Design: 2008
Construction: 2008.4-2010.10
GFA: 26,856m^2
Floors: 3
Height: 39.06m
Award: The Second Prize of National Excellent Engineering Exploration and Design Award, 2011
The Second Prize of Excellent Engineering Design Award of Guangdong Province, 2011

1 总平面图
Site plan

2 自行车馆造型设计概念为亚运会五羊会徽与极限运动头盔形象的融合
The building shape is envisioned by integrating the elements of five-ram emblem of Asian Games and x-sports helmet

3 自行车轮滑极限运动中心园区包括自行车馆、轮滑场、极限运动中心
Guangzhou Velodrome consists of cycling center, roller skating field and x-sports center

4 航拍图，金属屋面达到设计完成度
As shown in the aerial photo, the metallic roof is realized to the desired design effect

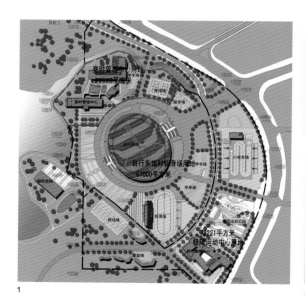

3

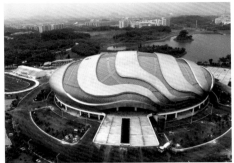
4

Sports Buildings 体育建筑 181

5

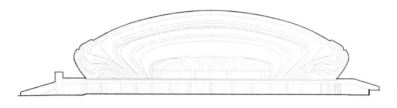

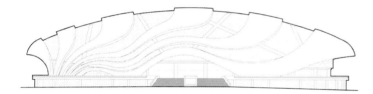
6

7

5 在自行车馆赛道上共产生了6项新亚洲纪录,为刷新纪录最多的比赛项目
  This track witnessed 6 new Asian records, dwarfing other competition events

6 立面图
  Elevation

7 屋面结构与局部双层的单层网壳结构,是目前国内超限大跨度结构的首例
  The single-layer latticed shell with partial double-layer for the roof is the first of its kind among the over-the-code large span structures in China

Sports Buildings 体育建筑

佛山市岭南明珠体育馆
Lingnan Pearl Gymnasium, Foshan

佛山体育馆是佛山市迎接2006年广东省第十二届运动会的核心建设项目之一，并于2010广州亚运会期间成功承办了拳击比赛。体育馆由三个多重圆水平环屋顶联体构成的建筑造型通透且独特，富有张力。夜晚由体育馆内透的灯光使建筑变得更为绚烂多彩，使体育馆建筑成为佛山城市的一颗璀璨的"岭南明珠"，塑造出欢快、开放、动感的城市景观，体现了佛山城市与体育产业的蓬勃发展。

体育馆位于佛山市禅城区季华五路，总建筑面积为73354平方米，包括8464座主体育馆一座，练习馆及大众馆等健身娱乐、商店、餐饮相关附属设施。三馆通过三个钢结构穹顶连为一体，结合部设置共享入口大厅，使建筑形态保持统一感。主赛场为70米×50米大小的长方形空间，可移动座席收藏在四边，当座席全部打开时，场地的有效尺寸为48米×33米，可进行篮球、手球、排球、网球等比赛。主体育馆设计时充分考虑了满足承办各种博览会、文艺演出、时装表演等活动的需求，各馆及附属空间在赛后得到了较为充分的利用，运营情况良好。为适应"全民健身"的时代需求，在用地西侧设置市民健身广场，健身广场通过连廊与三个馆相连，既保持了建筑整体良好的空间联系，也为市民提供了环境舒适的体育休闲娱乐空间。

该体育馆体现了诸多被动式环保、节能设计理念。建筑环状的层叠式屋顶将雨水导入首层的水池、绿化中，改善了环境小气候。屋面采用绝热材料以及隔声材料，侧向天窗通过屋檐的反射使进入室内的自然光较为柔和，避免产生眩光影响。电动开启窗可加强室内自然通风。设置的多个窗井为地下室提供了自然采光，达到有效节能的目的。

As one of the key projects developed for the 12th Guangdong Provincial Games in 2006, the Gymnasium successfully hosted the boxing event for 2010 Guangzhou Asian Games. The building composed by three multiple horizontal ring roofs looks transparent, unique and highly stretchable. At night, the Gymnasium lit from inside appears like a dazzling pearl, which constitutes an integral part of the lively, open and dynamic cityscape and demonstrates the booming urban development and sports industry of Foshan.

Located on Ji Hua Wu Lu, Chancheng District, Foshan, the Gymnasium with GFA of 73,354m² is composed of main arena, training gym, public gym as well as fitness, recreation, stores, F&B and other auxiliary facilities. Three arenas are connected through three steel domes and share one lobby at the connection to ensure the coherent building appearance. The main arena is designed into a rectangular space sized 70m X 50m, while the effective size is 48m×33m, which can host basketball, handball, volleyball and tennis games when the movable seats concealed around are fully unfolded. As the design for main arena takes into full account the demands for various expos, art shows, fashion shows and other events, the post-game use and operation of the arenas and auxiliary spaces are successful. In response to "mass fitness" initiative, a public fitness square is provided in the west and connected with the three arenas via bridge to maintain the favorable spatial connection and offer pleasant sports, leisure and recreation spaces to the public.

The Gymnasium also sees extensive use of passive environment protection and energy efficiency concepts. The annular cascaded roof guides the rainwater to the ground floor's pool and greening to improve the micro-climate. The roof is made of thermal and sound insulation materials, while the side windows soften the daylight reflected by the eave to avoid the glare. The motor-driven operable windows improve the interior ventilation, while the daylight shafts introduce the daylight into the basement for less energy consumption.

1

项目地点：广东省佛山市禅城区季华五路
设计时间：2003-2004
建设时间：2006
建筑面积：73354m²
建筑层数：地下1层，地上4层
建筑高度：36m
合作单位：日本株式会社环境设计研究所（EDI）
　　　　　日本构造设计集团（SDG）
曾获奖项：2007年度广东省优秀工程勘察设计奖一等奖
　　　　　2008年度全国优秀工程勘察设计行业奖建筑工程二等奖

Location: Ji Hua Wu Lu, Chancheng District, Foshan City, Guangdong Province
Design: 2003-2004
Construction: 2006
GFA: 73,354m²
Floors: 1 underground, 4 aboveground
Height: 36m
Partner: EDI + SDG
Awards: The First Prize of Excellent Engineering Exploration and Design Award of Guangdong Province, 2007
The Second Prize of Architectural Engineering under National Excellent Engineering Exploration and Design Award, 2008

1 在外部形态上,佛山体育馆主副各馆与健身广场均为圆形,并呈虚实互补关系,连廊和水池环绕它们一周更增强了整体感
  The circular main arena, auxiliary gym and fitness square are complementary to each other in solidness and void. The bridge and pool around enhance their integrity

2 从入口广场看体育馆建筑群体,其流线型穹顶结构给人留下独特而深刻的印象
  Viewed from the entrance square, the arena complex is highly impressive with the streamlined dome

3 剖面图
  Section

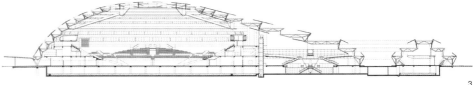

Sports Buildings 体育建筑 185

4

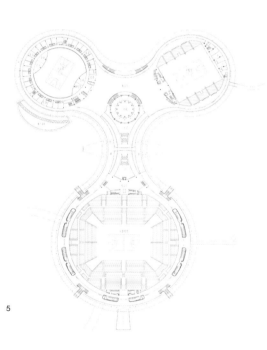

5

6

4 屋面采用连续的穹顶钢结构,其特色在于引进了斗拱的概念,强调了水平环桁架的作用,斗拱式穹顶网壳结构技术是首次在大跨度空间结构上应用。以短柱桁架和斜杆支撑水平环而取得平衡,这充分体现了一种继承与创新、建筑与结构完美结合的设计思路
The roof is characterized by continuous steel dome and concept of "Dougong" (bucket arch) to underline the effect of horizontal ring truss. The latticed shell technology of Dougong-supported dome is firstly applied in large span spatial structure. The balance realized by short column truss and diagonal horizontal ring fully demonstrates the perfect integration of legacy and innovation, architecture and structure

5 平面图
Plan

6 健身回廊围合的绿化广场塑造了良好的城市景观,回廊也是重要的群众健身场所
The green square framed by fitness corridors contributes to attractive cityscape, while the corridors offer place for public fitness

7 主赛场为70米×50米大小的长方形空间,可进行篮球、手球、排球、网球等比赛,以及举办各种博览会、文艺演出、时装表演。
The main arena is designed into a rectangular space sized 70m × 50m, which is available for basketball, handball, volleyball and tennis games as well as expo, art show and fashion show

Sports Buildings 体育建筑 **187**

惠州市金山湖游泳跳水馆
Jinshan Lake Swimming and Diving Natatorium, Huizhou

惠州市金山湖游泳跳水馆是广东省第十三届省运会场馆之一，设有满足国际标准赛事及残运赛事的游泳池、跳水池、热身训练池、训练室及室外训练池。它在非赛事阶段对公众开放，并作为当地运动员日常训练场地，以其鲜明新颖的建筑形象和极具感染力的室内空间，为市民提供了一处现代体育与文明科技高度结合的运动场所，为城市营造了一处具有标志性意义的公共活动场所。设计以"山水意象"为主题，阐述环境和建筑的关联合奏的意义，巧妙利用丘陵、水与山谷地状特殊性，建筑体布局呈平缓起伏状，在青山衬托下呈现出多姿多彩的表情。

游泳跳水馆既像一座座起伏的山峦，又如运动员在水中搏击的阵阵波浪，这一设计理念不仅体现在其流动的屋顶造型上，更与该体育建筑的功能和空间的需求相配合。游泳跳水馆内跳水池、比赛池、训练池一字排开，在功能上比赛与训练的区域分开，而空间相连，使馆内的空间更显宏大。流动的屋顶造型，营造出富层次感与趣味性的场馆内部空间，三维造型的跳台及椭圆形的玻璃背景板设计是馆内空间的焦点，运动员从里面走出来，就像登上一个竞技的舞台，一个人生的舞台，一个开启梦想的舞台。

设计采用自由的建筑形态，在形体交接处理上采用非线性手法，强调钢结构的韵律美。通过组合变形手法和材质对比处理，使形体简洁纯净又不乏细部。场馆屋盖结构纵向和横向均呈流线型，钢结构屋盖纵向高度变化较大，为减小风及温度的纵向作用对结构造成的不利影响，结构设计借鉴桥梁设计的经验，通过合理地选用固定支座、单向活动支座以及多向活动支座等新型抗震减振支座，释放了温度作用，减小了结构构件的尺寸，既经济又满足了建筑师对建筑整体效果的要求，使得室内结构看起来更为轻巧、飘逸。

As one of the venues for the 13th Guangdong Provincial Games, Jinshan Lake Swimming and Diving Natatorium contains the swimming pools, diving pools, warming-up pools, training rooms and outdoor training pools for standard international events and the Paralympics. During the non-competition period, the Natatorium is open to the general public and serves as daily training place for local athletes. With distinctive architectural appearance and impressive interior space, the Natatorium offers a sports venue that perfectly integrates modern sports with S&T and creates a representative urban space for public activities. Themed on picturesque landscape, the design proposal intends to elaborate the concert between the environment and the building. By following the terrains of hills, water and valleys, the building gently undulates, presenting diversified appearances against the mountains.

The building resembles undulating hills or waves splashed by competing swimmers. Communicated via the floating roof, such design concept further meets the functional and spatial demand of this sports building. The diving pool, competition pool and training pool are lined up to concurrently realize the functional separation between and spatial connection of the competition and training area, thus bring about a larger space. The floating roof generates hierarchical and interesting interior space. The unique 3D diving tower and oval glass background become the focus of the interior space, from which the athlete will embrace the stage of competition, life and dream.

With a free architectural form, the project features a non-linear approach at the building connection while highlighting the attractive rhythm of steel structure. Through combination, variation and contrast of materials, the building presents concise form and rich detail. The roof is streamlined vertically and horizontally. To minimize the adverse impact on the structure brought by the longitudinal effect of wind and temperature due to the considerable height variation of steel roof, the structural design references the bridge design approaches to reasonably use innovative anti-seismic and vibration damping support including fixed support, one-way movable support and multi-directional movable support as a way to release temperature effect and downsize the structural components. This cost-effective way contributes to the desirable architectural effect and lighter and more graceful interior structure.

项目地点：广东省惠州市
设计时间：2006-2007
建设时间：2006-2010
建筑面积：24574m²
建筑层数：3层
建筑高度：28.8m
曾获奖项：2006年国内竞赛第一名
2011年第六届中国建筑学会建筑创作佳作奖
2011年全国优秀工程勘察设计行业奖二等奖
2011年广东省注册建筑师优秀建筑佳作奖
2011年度广东省优秀工程勘察设计奖二等奖
2011年惠州市优秀工程设计一等奖

Location: Huizhou City, Guangdong Province
Design: 2006-2007
Construction: 2006-2010
GFA: 24,574m²
Floors: 3
Height: 28.8m
Awards: The first place of domestic competition in 2006
Excellent Award of the 6th ASC Architectural Creation Award, 2011
The Second Prize of National Excellent Engineering Exploration and Design Award, 2011
Excellent Architecture Creation Award by Guangdong Chapter of Association of Chinese Registered Architects, 2011
The Second Prize of Excellent Engineering Design Award of Guangdong Province, 2011
The First Prize of Excellent Engineering Design of Huizhou City, 2011

1 总平面图
Site plan

2 设计充分发挥金属屋面系统的优势，墙面、屋面一气呵成
The advantage of metallic roofing system is given full play to form coherent walls and roof

3 游泳跳水馆正立面
Natatorium's front elevation

4 金山湖游泳跳水馆造型柔和起伏，与周边环境相协调
Soft and undulating Natatorium in concert with surroundings

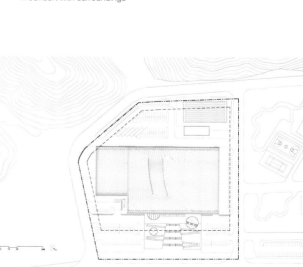

3

4

190 体育建筑 Sports Buildings

5/6 游泳馆的内部结构强调了钢结构的韵律美和简洁美，利用规则的结构体系创造出自由变化的三维曲面空间
The interior structure highlights the rhythm and concision of the steel structure with regular structural system to create a changing three-dimensional curve space

7 跳台背景墙设计独特，给运动员提供一个展示实力的"舞台"
The unique background wall of the diving tower offers athletes a show stage

8 在屋面高低错动之处设有侧窗，为室内引入自然光线
Side windows at different roof height introduce daylight

广州市花都区东风体育馆
Dongfeng Gymnasium, Huadu District, Guangzhou

"青山翠微迎露珠，秋意新雨霁绿藓。" 花都区东风体育馆采用简洁圆润的椭圆形，与其西南侧的康体公园和谐共生，与青山相互映衬，以独特的建筑形象与空间特点成为当地地标。本项目是2010年广州亚运会期间新建的体育场馆之一，也是花都中心城区西部的文化体育中心。体育馆能容纳8000名观众。满足体育场馆赛时及赛后的多功能使用是建筑总体设计的核心价值。

在权衡业主的需求、资金、施工周期等综合因素后，我们放弃了非线性的自由建筑形态，以简单实用为基调，采用最纯粹的几何形体量处理方法，以完整的体量与开阔的空间和周边建筑取得平衡。在椭圆体型基础上，金属屋面板、玻璃幕墙以及玻璃雨棚共同描绘出飞扬动感的曲线，整体具有自由流畅的动态形象。不规则的侧窗、天窗在打破简单外观形象的同时，为室内空间带来灵动、通透的光影。从第五立面看，体育馆又成了含苞待放的花蕾。体育馆可以概括为两大功能空间：比赛馆和训练馆。主从有别的两个体量由曲线平台连接，形成自由、流畅的统一整体。比赛馆内观众休息厅与比赛大厅相互联通，在有限的结构空间中实现建筑使用空间的最大化，创造出扩大化的视觉感观及使用效率，从而节省面积。结构设计借用了"箍桶原理"，在国内首次创造性地设计出了环形管内预应力大跨度钢结构体系。其结构构件小，安全性高，是同类体育场馆用钢量最低的，最大限度地节约了投资造价。

在满足大型国际体育赛事的复杂使用要求的同时，充分考虑全民健身需求及不同公众活动的需求。亚运后，场馆成为该区域的全民健身运动中心，也作为举办各种大中型活动（演唱会、展会）以及汽车城工业产品的展销平台等得以有效利用。

项目地点：广州市花都区
设计时间：2008
建设时间：2008-2010
建筑面积：31416m²
建筑层数：4层
建筑高度：33.3m
曾获奖项：2008年国内竞赛第一名
广东省第六次优秀建筑佳作奖
2011年度广东省优秀工程勘察设计奖二等奖
2011年度广东钢结构金奖"粤钢奖"
2011第十四届中国室内设计大奖赛学会奖
2011年度全国优秀工程勘察设计行业奖三等奖

Inspired by the poem that goes "*The emerald mountains joyfully welcome morning dews while the autumn rain hastens the growth of mosses*", Dongfeng Gymnasium takes the shape of a concise and mellow ellipsoid to harmonize with the fitness park to the southwest and respond to the emerald mountains. As one of the newly built sports venues for 2010 Guangzhou Asian Games, the Gymnasium with unique architectural image and spatial features serves as the local landmark and cultural and sports center in the west downtown of Huadu. The master plan of this 8,000-seat gymnasium aims to ensure the multi-functionality during and after the game.

Considering the client's demands, costs, construction period and other factors, we decided to divert from the non-linear free form; instead adopt simple geometric volume and take clear shape and efficient functionality as design guidelines to harmonize with the surrounding buildings through complete volume and open space. Based on the olive shape, the contours generated by metallic roof, glass curtain wall and glazing canopy compose the freely flowing and dynamic image. The irregular side window and skylight diversify appearance and bring the interior space interesting play of light and shadow. Viewed from the 5th façade, the Gymnasium looks like a budding flower. The Gymnasium consists of two functional spaces, i.e. primary competition arena and sub-ordinated training arena, which are connected by curved platform to form a free and smooth unity. The spectator lounge is connected with the competition hall in the competition arena as a way to maximize the usable area within the limited structural space, expand the space visually and enhance the efficiency. By referencing the "wooden barrel principle", the structural engineer creatively proposes the pre-stressed large span steel structure system in ring tube for the first time. Such system featuring small components, high safety and lowest steel consumption in similar sports venues realizes the maximum cost effectiveness.

While the requirements of major international competitions are met, the demands of public fitness and other activities are also considered. After the Asian Games, apart from being a district public fitness center, the Gymnasium will also be used to host various large and medium sized events (concert and exhibition) and trade fair for the industrial products of the Auto City.

Location:	Huadu District, Guangzhou
Design:	2008
Construction:	2008-2010
GFA:	31,416m²
Floors:	4
Height:	33.3m
Awards:	The first place of domestic competition in 2008
	The Sixth Excellent Architecture Creation Award by Guangdong Chapter of Association of Chinese Registered Architects
	The Second Prize of Excellent Engineering Design Award of Guangdong Province, 2011
	Gold Award for Steel Structure Design in Guangdong (Yuegang Award), 2011
	Academy Award of the 14th China Interior Design Competition, 2011
	The Third Prize of Excellent Engineering Survey and Design Award in China's Engineering Survey and Design Industry, 2011

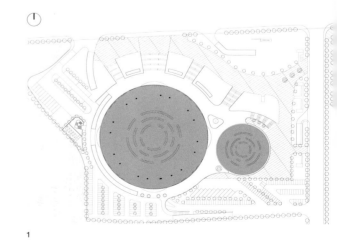
2

1

1 总平面图
　Site plan

2 体育馆采用简洁圆润的椭圆形,如两颗露珠在落在自然景观之中
　Concise and mellow ellipsoid like two beads of dew within the natural landscape

3 体育馆概念手稿
　Sketch of conceptual design

4 体育馆南立面
　South façade

5 体育馆建造过程模拟
　Construction process simulation

6

7

8

6	体育馆夜景 Night view
7	体育馆局部设有尺度适宜的下沉庭院 Properly scaled sunken court
8	体育馆门厅 Foyer
9	通向后勤服务区的通道及二层平台的大台阶 Access leading to BOH and large steps of F2 platform

肇庆新区体育中心
Zhaoqing New Area Sports Center

肇庆新区体育中心，作为第十八届广东省运动会的唯一新建场馆，其主要由体育馆（8000座体育馆主馆和训练馆）、专业足球场（20000座）及相关配套设施组成。肇庆新区体育中心集约布局，空间互相渗透，整体设计，结合赛后商业开发营运，创造一个开放、公共，面向社区、城市的全民健身中心。

体育中心以一个流畅优雅的金属屋面来统领多个场馆，兼顾建筑整体和谐统一同时场馆间又营造出大片连贯舒展的灰空间，将市民的活动巧妙地融入其中，营造了积极开放的城市空间。专业足球场突破性地采用半开放式设计，金属屋面向河岸打开，整体屋面形成一个不闭合的环，结合南侧开放看台，使球场与公园连成一体。场内观众可看到肇庆的城市山水，场外观众也可感受比赛的热烈气氛。足球公园与北侧的体育中心紧密相连，公园用地主要设置1个带跑道的室外体育场、2个小型室外足球场、8个室外篮球场、6个室外网球场及公园配套设施等，作为体育中心项目的辅助设施，在省运会期间发挥重要补充作用。

赛后运营是本次设计的重点之一。体育场馆的部分功能用房赛后将改造为商业空间，位于体育馆三层的贵宾休息厅也将改造为湖景餐厅。城市客厅尺度宜人，利于赛后商业气氛营造。集约、可持续、可实施的专业体育场馆，设施完善、尺度适宜的足球公园，以及多层次立体滨河空间紧密结合，成就一个面向肇庆新区的标志性建筑，一个服务周边社区的城市公共空间以及全民健身中心。

As the only venue newly built for the 18th Guangdong Provincial Games, the Sports Center is composed of a gymnasium (8,000-seat main gymnasium and training gym), 20,000-seat professional football stadium and supporting facilities. With the compact layout, interpenetrated spaces and integrated design in view of the post-game commercial development and operation, the Sports Center serves as an open, public, and community-oriented urban fitness center.

A smooth and elegant metallic roof unifies various facilities, creating harmonious and coherent building appearance while offering generous expanse of grey spaces for public activities and dynamic open urban space. The professional football stadium adopts innovative semi-open design, where the metallic roof opens toward the river as an unenclosed annular ring to connect to the park with the help of the open stand in the south. This way, the spectators inside the stadium may enjoy the picturesque view, while those outside can feel the exciting atmosphere of the game. Closely connecting to the Sports Center to the north, the football park is designed with 1 outdoor stadium with track, 2 small outdoor football courts, 8 outdoor basketball courts, 6 outdoor tennis courts and supporting facilities. As the auxiliary facility of the Sports Center, the park played an important complementary role during the Provincial Games.

The post-game operation is made one of the priorities in the design. Some functional rooms will be renovated into commercial facilities, while the VIP lounge on F3 will be transformed into river-view restaurant after the game. The human-scaled urban space makes for the post-game commercial ambience. The compact, sustainable and viable professional gymnasium, full-fledged and human-scaled football park and the multi-layer riverfront space jointly contributes to this iconic building in the New Area of Zhaoqing, also a pubic urban space and fitness center serving the neighborhoods.

项目地点：广东 肇庆
设计时间：2015-2016
建设时间：在建
建筑面积：85000m²
建筑层数：局部地下1层，地上4层
建筑高度：48m
曾获奖项：2015年国内投标第一名

Location: Zhaoqing, Guangdong Province
Design: 2015-2016
Construction: ongoing
GFA: 85,000m²
Floors: 1 underground partially, 4 aboveground
Height: 48m
Award: The first place of domestic bidding, 2015

1

1　总平面图
　　Site plan

2　鸟瞰图
　　Bird's eye view

3　从长利涌望向体育中心，专业足球场屋面向河岸打开
　　View from Changli Canal: roof of football stadium opens toward the river

4　专业足球场和体育馆面向城市道路，各有相对独立的出入口广场
　　Facing the urban road, the professional football stadium and gymnasium are provided with separate entrance squares

5　足球公园设有各种运动、休闲设施，是城市公共生活中心
　　The football park with various sporting and leisure amenities is an urban center for public life

惠州博罗县体育中心体育场
Sports Center Stadium, Boluo County, Huizhou

惠州博罗体育场作为2010年广东省第十三届运动会比赛场地之一，是国内首次采用斜拉索+桅杆+大跨度悬臂钢桁架结构的体育场。

博罗体育场结构形式新颖、传力途径清晰、结构刚度大、构件简单、结构便于施工。整个结构犹如人体一样拥有多个可大角度转动的关节，风力大时位移大，风速小时静止。在设计阶段进行了仔细的几何非线性分析，通过调节拉索的截面、初拉力和结构的预拱度，控制主桁架的竖向位移，使整个结构受力最优。体育场看台的总体形态是一个倾斜的、呈新月状的楔形体，该形体是呈上浮的，具有升起感。这种形体似乎预示着古城浮出水面。结构体系采用每榀斜撑结构富有力度感，体现了向上的、朝气蓬勃的精神面貌。

博罗体育场设座位约15000个，分东西两个看台，看台下部为钢筋混凝土结构，屋盖为大跨度悬索桁架结构，顶面覆盖PVC张拉膜。西看台混凝土部分最高处高度约为22.5米，顶盖最大悬臂跨度为39.1米。东看台混凝土部分最高处高度约为17.6米，顶盖最大悬臂跨度为30.1米。

As one of the venues for the Thirteenth Guangdong Provincial Games in 2010, the project is the first Chinese stadium structured by stay cable + mast + large span cantilever steel truss.

The structure is featured by innovative form, clear load-bearing path, high stiffness simple component and easy construction. Similar to a human body, the structure comes with multiple highly rotatory articulations allowing for different displacements varying with wind. With cautious geometric non-linear analysis in design, the overall structural load-bearing is most optimized by adjusting the cable section, initial tension and structural pre-camber and controlling the vertical displacement of main truss. Taking the shape of inclined crescent wedge, the stadium stand ascends, suggesting the rising of this historical city. The powerful structure with diagonals for each truss frame is an inspiring display of the sportsmanship.

The 15,000-seat stadium is designed with east and west stand based on reinforced concrete structure and covered by the large-span cable truss roof and PVC tensioned membrane ceiling, where the highest concrete structure is 22.5m (west) and 17.6m (east) and the largest cantilever span of the roof is 39.1m (west) and 30.1m (east).

2

项目地点：惠州博罗县
设计时间：2009
建设时间：2009-2013
建筑面积：18621.9m²
建筑层数：地上3层
建筑高度：28.95m
获得奖项：2015年度全国优秀工程勘察设计行业奖建筑结构二等奖
　　　　　2015年度广东省优秀工程勘察设计奖建筑结构专项二等奖
　　　　　2015年度广东省优秀工程勘察设计三等奖

Location: Boluo County, Huizhou City
Design: 2009
Construction: 2009-2013
GFA: 18,621.9m²
Floors: 3 aboveground
Height: 28.95m
Awards: The Second Prize of National Excellent Engineering Exploration and Design Award - Building Structure, 2015
The Second Prize of Excellent Engineering Exploration and Design Award of Guangdong Province - Building Structure, 2015
The Third Prize of Excellent Engineering Exploration and Design Award of Guangdong Province, 2015

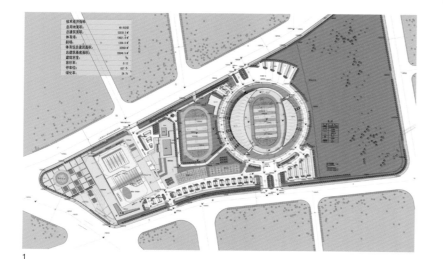

1

3

1. 总平面图
   Site plan

2. 体育场看台的总体形态是一个倾斜的，呈新月状的楔形体
   The stadium stand takes the shape of inclined crescent wedge

3. 整个结构犹如人体一样拥有多个可大角度转动的关节
   Similar to human body, the structure comes with multiple highly rotatory articulations

4. 建筑罩棚采用山峦起伏的形态，体现惠州罗浮山和体育运动结合的特点
   The cantilevered roof resembles the rolling hills and incorporates the elements of Luofu Mountain and sports

5/6 立面图
    Elevation

医疗建筑

没有比医院更令人有感触的建筑了，面对病痛、衰老、死亡，医院天然充满负能量，为了保证质量、效率、流程的"医疗器械"状态，充满着繁文缛节的规范限制；也没有比医院更复杂的建筑了，我们设计的不只是医院，还是流程、体验和结果。随着医疗保健理念从传统医学向现代医学转变，从只关注引发疾病的本源（生物医学）过渡到多角度至方面的健康观（生物——心理——社会医学），强调有益康复的环境与自然健康心理对病患的积极影响。

虽然传统医院，或一个标准的医院，就是聚集治疗的场所；是在医疗工艺的严格限制、医疗流程的严格管控下，以医技为绝对核心的程序化设计，如医疗器械一样精密、高效运行；"人"，仅仅是个生理符号。但现在是一个人人对自己的健康负责的时代，是一个人人渴求信任和信仰的时代，人人都有一个梦想："通过明确的指引，快速走进干净整齐的空间；在井井有条的环境下，医护人员认真处理着我们的诉求，并得到医护人员的全程护理，顺利的渡过生命中最脆弱或最快乐的时光"。随着人的信仰和对生命的尊重（以病人为中心）等现代医疗观念的确立，医疗建筑越发注意空间的营造，使其具备家的温馨，同时充满艺术气息。

广东省建筑设计研究院65年来的医疗建筑实践，从最早广州市第一人民医院到最新的广州市妇女儿童医院南沙分院，基于对人的生命和信仰的敬畏，一直保持着与世界潮流接轨的前沿医疗建筑设计理念。运用John Weeks的indeterminate Architecture理论，通过新陈代谢、细胞式生长的方式改造1956年建设的中国石油中心医院(廊坊)，让老医院重新焕发青春；运用"Best Buy"、"Harness"、"Nucleus"的理论，成功地在株洲医院、重庆儿童医院的设计实践中采用标准化模块的设计手法，并采用了"Oxford Method"的方式工业化建造，实现资金、质量和建造效率的可控性。

"深入研究才能精湛设计"，不断进取的GDAD的医疗设计团队经历了梅州市人民医院、株洲市中心医院、惠州市妇幼保健院和广州市妇女儿童医院南沙分院等多个大型三甲医院的全建设过程，提供方案到装修施工、工艺到建筑全过程的设计服务，充分掌握并熟练运用巨相相依、互相制约的医疗工艺、医疗单元之间的逻辑关系，发展出"Nucleus"核心医院模式，将运营费用及使用者的感受纳入考虑，通过"机变建筑"的设计手段，不断创作出适应现在及将来、社会和政治价值观影响下的医疗建筑。

GDAD在株洲市中心医院和重庆儿童医院等超大型医院的设计中，首次采用交通枢纽建筑的设计手段来解决医疗建筑的流线问题，通过医疗街组织交通，使医疗建筑流线井井有条，也有了空间开阔的大厅及不受干扰的等候空间；并通过对医疗工艺充分理解、尊重，根据用地条件和使用需求，衍生多种单元组合模式，使严谨的医疗工艺与高品质的建筑空间和谐共生；在广州市妇女儿童医院南沙分院的设计中，首次通过对未来科技发展的了解和循证医学的数据分析，模块化组合医疗功能，使医疗空间产生了组合和变化、并能有序生长；使用绿色、节能的建筑技术，使灿烂的阳光、清新的空气、宜人的绿色庭院进入封闭的空间，给医疗建筑带来家的温馨，带来心灵的慰藉；智能化技术的全面应用，使医务人员实现远程监控、有序管理，让流程如医疗器械一样精密运行；充分使用自动物流系统，使医疗建筑也有了像城市一样的高效管廊系统，保证建筑在全寿命周期的顺利运行。

为创造充满理想和激情的现代医疗建筑，GDAD一直在理性和创新中探索，游走于理性与浪漫、精密与温馨、开放与私密，管控与交流等矛盾中，为实现每个人的梦想而前行。

Medical Buildings

There is no other building as emotional as hospital that upsets and frustrates people with illness, aging and death and is fraught with regulatory red tape to ensure the quality, efficiency and process in a state-of-control. The hospital is of the highest complexity, as the design comes to the process, experience and result apart from the physical structure. Now that the conventional healthcare concept is shifted toward a modern one, a multi-angle and all-round view of health (biomedicine - psychological medicine - social medicine) is developed out of the previous simple concern about etiology (biomedicine), which values the positive impact on the patient by the favorable rehabilitation environment and natural and healthy psychology.

Designed as a place for centralized medical treatment absolutely centering on medical skill, a conventional or typical hospital operates like accurate and efficient medical device under the strict control of medical technology and process, where "human" is only a physical symbol. However, today everyone is supposed to take care of his/her own health and longs for trust and faith. It is long-expected that we can be clearly guided to easily access a clean and neat space where we can be properly treated by the medical staff in a well-organized environment and spend our frailest or maybe the happiest time in our life. With the establishment of modern medical concept that shows respect to human's faith and life (patient-centered), the more emphasis is given to creating home-like and artistic spaces in medical buildings.

Our medical building design practices in the past 65 years ranged from the earliest Guangzhou No.1 People's Hospital to the latest Nansha Branch of Guangzhou Women and Children's Hospital. Out of the reverence for human life and faith, we have been adhering to the cutting-edge medical building design concept that keeps abreast with the international trend. We employed John Weeks's concept of Indeterminate Architecture in renovation of CNPC Central Hospital (Langfang), which was built in 1956, and revived the old hospital through metabolism and cellular growth. We also applied the concepts of Best Buy, Harness and Nucleus in the standard modular design for Zhuzhou Hospital and Chongqing Children's Hospital, and properly controlled the cost, quality and construction efficiency through the industrialized construction of Oxford Method.

Quality design comes from in-depth study. Following the full-process development of the People's Hospital of Meizhou, Zhuzhou Central Hospital, Huizhou Women's and Children's Healthcare Hospital, Nansha Branch of Guangzhou Women and Children's Hospital and other 3A hospitals where GDAD provided the full-range design services from design scheme to interior construction, from technical process to architecture, our medical building design team has fully understood and gained proficiency in the logic relation between the interdependent and complementary medical procedures and units; moreover, developed the Nucleus hospital mode. This mode takes into account the operation cost and user's experience, and through the design approach for Indeterminate Architecture, creates the medical building resilient to present and future social and political values.

In the design of Zhuzhou Central Hospital, Chongqing Children's Hospital and other super large hospitals, we for the first time addressed the circulation of medical building with reference to the design approach of transportation hub. The traffic organized by medical street brought well-organized circulation, spacious lobby and undisturbed waiting area; based on the full understanding of and respect for the medical procedures, site conditions and functional demands, multiple unit combinations were produced, realizing the harmonious coexistence between the meticulous medical procedures and quality building space. In the design of Nansha Branch of Guangzhou Women and Children's Hospital, we for the first time conducted the modular combination of medical functions based on our understanding of the future technology development and the data analysis of evidence-based medicine, allowing for the combination, variation and orderly growth of medical spaces. We also employed the green and energy efficient building technologies to introduce the sunlight, fresh air and pleasant green courtyard into the enclosed space, fostering a cozy and warm home-like atmosphere in the medical spaces. The full application of the intelligent technologies enabled the medical staff to conduct remote monitoring and proper management, ensuring the same precise medical procedures as the medical equipment. The automatic logistic system provided the efficient utility tunnel to the medical building, just like those used in the cities, and, ensured the smooth building operation during its full life cycle.

To create modern medical buildings of both rationality and emotion, We have been exploring between conflicting elements like rationality vs. emotion, accuracy vs. coziness, openness vs. privacy, control vs. communication, etc, forging ahead to fulfill our dream.

中山大学肿瘤防治中心
Sun Yat-sen University Cancer Center

中山大学（原中山医科大学）肿瘤防治中心医疗科研楼是当时国内最大型的门诊、医技、住院、教学、科研、办公、后勤等几乎所有专科医院功能于一体的单幢高层建筑。坐落于广州市东风路、执信路及先烈路交汇处，地下两层，地上二十三层，日门诊量1500人，病床数900床。工程采用集中式布置，有效节约用地，最大限度地减少施工过程中对医院正常运作的影响。

项目设计采用了医院大堂酒店化、自动扶梯应用于门诊交通、大范围采用轨道式物流系统、全楼宇智能化系统、病房空调系统过滤优化、轻质墙体隔断以及装修细节和标识等当时少有的先进设计理念和技术，采用符合医疗建筑行为习惯的导向明确便捷、洁污分开、内外分开的流线设计和满足医疗建筑使用要求、设备要求的功能分区布局，为提高医院效率、人性化感受和有效地发挥集中式医疗建筑的优势做出来有益的探索和实践。

项目功能繁多、设备先进、管线复杂，涉及多个专业工种和医疗设备专业公司和厂家。建筑师从规划方案构思到二次装修施工全过程参与，使工程设计思想和整体效果得以落实和保证，也使得整个设计过程高效协调的运作。

Located at the junction of Dong Feng Lu, Zhi Xin Lu, and Xian Lie Lu in Guangzhou, the medical research building of Sun Yat-Sen University (formerly the Sun Yat-sen University of Medical Sciences) Cancer Center has 2 floors underground and 23 floors aboveground, receiving 1,500 outpatients daily and equipped with 900 hospital beds. The building features a centralized layout and was, at that time, the largest singular high-rise in China that integrated specialized hospital functions including outpatients, medical technology, inpatients, teaching, research, office, back of house etc. with efficient use of spaces and minimum impact on hospital operation during the construction.

The hospital is designed with advanced concept and technology rare at that time, including a hotel-style lobby, escalator for outpatients' circulation, rail-type logistics system, intelligent building system, ward AC system filtration and optimization, light-weight partition wall, decoration details and signage etc. The circulation system complies with the common behaviors and practices in a medical building, ensuring easy way-finding, the separation of clean area from contaminated area, and the separation of the inpatient area from the outpatient area. The functional zoning and layout meet the building's functional and EMP demands. So the project is an inspiring experiment on improving hospital efficiency and humanization and giving full play to the advantage of a centralized hospital building.

The hospital comes with multiple functions, sophisticated equipment and complicated pipelines, involving many disciplines, medical equipment suppliers and manufacturers. The architect was on board throughout the entire process from planning to fit-out to ensure the implementation of design concept and the overall effect, as well as the high efficiency and coordination during the design process.

项目地点：位于东风路、先烈路与执信路的交界处
设计时间：1996-2002
建设时间：2000-2002
建筑面积：83200m²
建筑层数：23层
建筑高度：94m
曾获奖项：2005年度全国优秀工程勘察设计行业奖三等奖
2005年度广东省优秀工程勘察设计奖一等奖
广东省注册建筑师协会优秀建筑创作奖

Location: Crossing of Dong Feng Lu, Xian Lie Lu and Zhi Xin Lu, Guangzhou
Design: 1996-2002
Construction: 2000-2002
GFA: 83,200m²
Floors: 23
Height: 94m
Awards: The Third Prize of the Ministerial-level Excellent Engineering Design Award, 2005
The First Prize of the Excellent Engineering Design Award of Guangdong Province, 2005
Excellent Architecture Creation Award by Guangdong Chapter of Association of Chinese Registered Architects

1 总平面图
  Site plan

2 与平面功能高度统一的建筑体型通过立面材料、颜色变化创造出典雅大气的建筑形象
  Building form is highly consistent with the planer functions, portraying a generous and elegant building image through the changes of façade material and color

3 富有韵律感的立面形式和细节处理，显示出一种毫不张扬的独特个性
  Rhythmical facade pattern and fine details are presented in a unique and modest manner

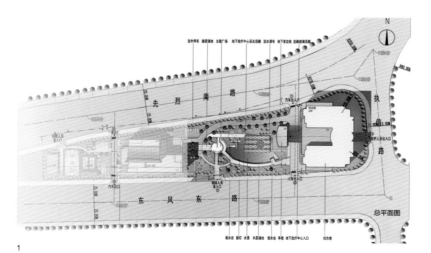

株洲市中心医院
Zhuzhou Central Hospital

株洲中心医院是一所拥有2300床大型综合医院，年门诊100万人，年住院5万人，总建筑面面积20万平方米。

中心医院布局方案的构想经过多方案的探讨，通过单元模块化沿医院街主轴扩展模式，实现了可持续发展，方案布局兼顾了传统水平式医院与绝对集中医院的主要优点，使得土地利用最大化，强调了资源整合和集中管理，开敞明亮的中庭和融入生态景观的医院街将门诊区、医技区及住院区有机串联，确保就诊流程简洁明了，医疗活动高效运行，为患者创造舒适宜人，高效便捷的就医环境。

建筑营造了亲切宜人的空间环境，减轻了病患等候时的焦虑和不安，创造出被接待的氛围。各功能空间具有高度的连续性和识别性，通过良好的引导措施减少患者的盲目流动和由此产生的焦虑情绪，更注意物理环境，包括光、声、热、空气质量、地面摩擦系数等。

Zhuzhou Central Hospital is a large general hospital with 2,300 beds, receiving an annual average of one million outpatients and 50,000 inpatients. The total floor area of the hospital is 200,000m^2.

Through much deliberation, the design of the hospital was finalized into a layout of module units extending along the main axis of the hospital street to realize sustainable development. With combined advantages of both traditional horizontal hospitals and fully centralized hospitals, land utilization is maximized with focus on resource integration and centralized control. The open and bright atrium and the hospital street, as part of the eco-landscape, connect outpatients, technology-aided diagnosis area and inpatients areas in an organized manner, ensuring the clearly defined service process and efficient operation of medical activities while creating a comfortable, efficient and convenient environment for patients.

The building offers a friendly and pleasant environment where patients are warmly received and their anxiety and stress relieved during the waiting. Functional areas are arranged continuously and can be identified easily with clear orientation. This helps reduce disorder in circulation and the resulting anxiety. The physical environment, including light, sound, heat, air quality and floor friction coefficient etc., has been well taken into consideration.

2

3

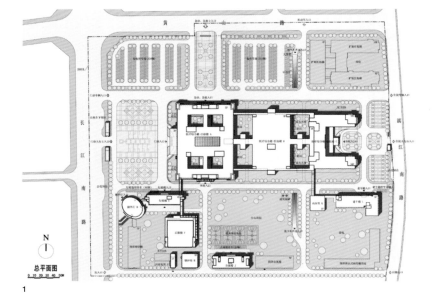

1

项目地点：湖南省株洲市
设计时间：2008
建设时间：2012
建筑面积：20万m²
建筑层数：16层
建筑高度：65.6m

Location:	Zhuzhou City, Hunan Province
Design:	2008
Construction:	2012
GFA:	200,000m²
Floors:	16
Height:	65.6m

1 总平面图
  Site plan

2-5 建筑细部用心雕琢，材质体块色彩的碰撞下反复推敲，达到美观，大方，怡人的效果
  Fine details of building achieved through deliberation of material, shape and color contrast to present an aesthetic, elegant and pleasant appearance

6 建筑群落完整方正，轴线明晰，气势恢宏，材质古朴，引人注目
  Square building, clear axial layout and primitive and simple materials jointly present a spectacular and attractive view

7 建筑在灯光的映照下典雅端庄，前广场尺度怡人，各个出入口位置合理
  An elegant and dignified building in lighting, with a well-proportioned front square and a rational layout of entrances

梅州市人民医院
The People's Hospital of Meizhou City

梅州市人民医院（黄塘医院）的前身是德济医院，由瑞士基督教巴色传道会派德籍医师韦嵩山博士于1896年来梅创办，距今已有100多年历史。现今是梅州市唯一的大型综合性国家三级甲等医院，广东省高等医学院校教学医院，中山大学教学医院，中山大学博士后流动站科研基地。

由于本次设计的两栋大楼（门诊综合大楼及住院楼）是在原有旧院区拆迁出来的不规则用地上兴建的，因此在总体规划布局上存在一些制约，院方希望得到的轴线关系难于体现。在吸取了梅州传统建筑围龙屋及"圆文化"的精髓下，本方案建筑平面采用了圆弧形布局，在总平面关系上形成了一个"S"形，并穿越了两幢建筑之间的小溪，既满足了院方对于建筑物对称的要求，也使两幢不同朝向、不同轴线关系的建筑找到呼应关系，理顺了原有院区较为混乱的平面关系。

除此之外，门诊综合大楼的设计也是一个难点。受场地及资金限制，该综合楼集门诊、急救及医技功能为一体，流线容易交叉，且在首层还必须解决儿童门诊及传染病门诊的入口，设计上有着诸多难点。本设计通过两个垂直交通体，巧妙将上述三大功能自然分隔，又相互联系，大大精简了楼电梯的配置，节省了空间和投资。在首层入口处增加了风雨连廊，既解决了病人下车遮阳挡雨的问题，又有效延长了对外的入口界面，解决了各类型病人入口分类的问题。

项目地点：广东省梅州市
设计时间：2002
建设时间：2005
建筑面积：30000m²
建筑层数：地上9层，地下1层
建筑高度：39m
曾获奖项：广东省第十二次优秀工程设计三等奖

Location: Meizhou City, Guangdong Province
Design: 2002
Construction: 2005
GFA: 30,000m²
Floors: 9 aboveground, 1 underground
Height: 39m
Award: The Third Prize of the Twelfth Excellent Engineering Design of Guangdong Province

The People's Hospital of Meizhou City (Huangtang Hospital), formerly known as Deji Hospital, was founded in 1896 by Dr. Hermann Wittenberg, a German doctor sent to Meizhou City by the Swiss Christian missionary society Basel Mission, and has a history of over 100 years. Now it is the only large state-run 3A general hospital in Meizhou City, and is the teaching hospital for medical institutions of higher education in Guangdong Province, the teaching hospital for Sun Yat-Sen University, and SYSU's mobile station for post-doctoral research.

As the two buildings (outpatients complex and inpatients complex) are built on an irregular site where former hospital buildings were demolished, there are certain restraints in master planning, and the axial layout the hospital desires is hard to achieve. Referencing Meizhou's traditional architecture the Hakka Round-dragon House and the "Culture of Roundness", we propose an arc layout in the shape of an "S" that goes across a brook between the two buildings, realizing the symmetry required by the hospital and making the two buildings of different directions and axes echoing each other. Thus a well-organized layout is achieved.

Besides, the design of the outpatients complex is also a challenge. Due to limited site and cost budget, the complex has to integrate functions of outpatients, emergency and technology-aided diagnosis, so circulations can easily conflict with each other, meanwhile, the entrances to pediatric outpatients and infectious disease outpatients must be placed on the first floor. All these pose great challenges in design. The solution is to, via two vertical transportation cores, naturally separate the said three functions from each other while maintaining the connection. This approach reduces the unnecessary elevators and save both space and cost. A weatherproof corridor is added to the entrance on the first floor, providing sun and rain shelter for incoming patients while also expanding the entrance area to provide separate accesses for different types of patients.

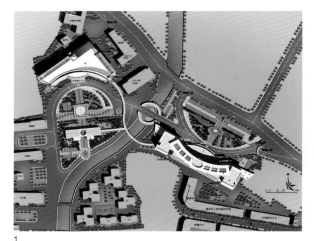

1 总平面图
  Site plan

2 室外现场低点效果图
  Exterior view from lower level

3 室外现场低点
  Lower exterior level

4　住院楼入口夜景
　　A night view of entrance to inpatients complex

5　门诊楼主入口低点视角
　　Lower level perspective of main entrance to outpatients complex

6　住院楼入口低点视角
　　Lower level perspective of main entrance to inpatients complex

7　住院楼主立面低点视角：圆弧造型结合对称的立面，简约中凸显细腻
　　Lower level perspective of inpatients complex's main facade: the façade integrating arc and symmetric pattern appears simple but exquisite

惠州市妇幼保健院
Maternal and Child Health Hospital of Huizhou City

惠州市妇幼保健院位于惠州市惠城区，规划用地面积3.72万平方米，总建筑面积56812平方米。建设总规模为500张床位，建筑总高33.9米。本项目由一栋八层综合主楼（门诊、住院、医技楼）、地下室、发热门诊及消化道门诊楼、餐厅及宿舍、地下污水处理站等部分组成。

建筑布局合理，流线清晰便捷，洁污分流明确，在满足安全、卫生和使用功能等基本要求的前提下，引入现代化医院的建设模式，采用高效节能的建筑布局，创造引人入胜的公共空间，营造舒适优美的室内外环境，体现时代气息的建筑造型，使其成为具有妇女儿童特色的现代化专科医院。建筑和环境设计注重园林化和生态化，采用了围合式庭院，将中心花园为空间核心，并与建筑周围的大小各式庭院穿插呼应，并采用多层次立体化的绿化处理。从而使各个部门都处于园林绿化的围绕之中，不但具有良好的采光和通风效果，同时也获得了优秀的景观效果。

采用绿色建筑的设计理念，运用多种设计手段、采用新材料、新技术，实现建筑节能和环保。采用新型能源设备——热泵，该设备可以利用空调热交换过程的余热以及空气热能为医院提供热源；采用生物降解的污水处理工艺，对医院污水先进行无害化处理，具有环保、节能等优点。

Located in Huicheng District of Huizhou City, the Maternal and Child Health Hospital of Huizhou City is planned with a site area of 37,200m^2 and a gross floor area of 56,812m^2. Equipped with 500 hospital beds, the 33.9m-high facility consists of an eight-floor complex (outpatients, inpatients and technology-aided diagnosis building), basement, fever and digestive outpatients building, canteen and dorm, underground waste water treatment plant etc.

The building features rational layout, easy and clearly-defined circulation and separation of clean circulation from the contaminated one. Apart from meeting the fundamental requirements of safety, hygiene and functionality etc., the project is also designed with attractive public spaces, comfortable and elegant interior and exterior environment, and contemporary building form thanks to the modern hospital development approach and energy-efficient building layout. As a specialized modern hospital for women and children, the project emphasizes the garden style and ecology in architectural and environmental design. For this, the enclosed courtyards are provided, with the central garden as the hardcore of the space. The central garden interweaves and echoes with the courtyards of varied sizes around the building, embracing all the departments with the multi-level greening. This not only contributes to favorable daylighting and ventilation effect but also offers attractive views.

Under the green building concept, a variety of design approaches, new materials and technologies are employed to realize an energy-efficient green building. For example, the heat pump, a new type of energy equipment, is used to supply heat for the hospital by reusing the waste heat and air heat from the air-conditioning thermal exchange; besides, the biologically degradable waste water treatment technology can conduct the non-hazardous pre-treatment of waste water from the hospital in an environmental friendly and energy efficient manner.

项目地点：惠州市惠城区河南岸
设计时间：2007.5
建设时间：2011.12
建筑面积：56812m^2
建筑层数：6-8层
建筑高度：33.9m
曾获奖项：2013年度广东省优秀工程勘察设计奖三等奖

Location: South bank, Huicheng District, Huizhou City
Design: 2007.5
Construction: 2011.12
GFA: 56,812m^2
Floors: 6-8
Height: 33.9m
Award: The Third Prize of Excellent Engineering Design Award of Guangdong Province, 2013

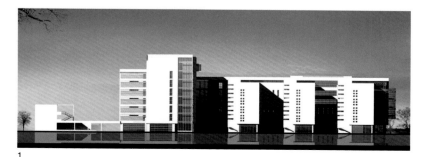

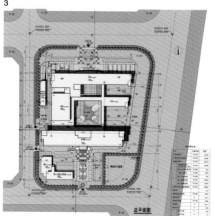

1 立面图
  Elevation

2 总平面图
  Site plan

3/4 门诊楼入口舒展大气，尺度宜人
  Generous and well-proportioned entrance to outpatients building

5 从城市界面角度出发，沿街立面设计错落有致
  Elegantly staggered street-front facades based on the urban edge

6/7 活泼的半围合室外中庭空间，有效改善了严肃的医疗环境
  Lively semi-enclosed outdoor atrium effectively relaxes the atmosphere of a medical environment

8 门诊大厅空间宽敞，给患者一个舒适的就医环境
  Spacious outpatients' hall offers a comfortable medical environment for patients

海螺医院
Conch Hospital

在安徽省芜湖市，海螺医院是拥有1500床规模的大型综合医院。总体规划中结合地形、周边环境、道路交通等各项因素，将医院分为以下五个功能区：门诊与医技区、住院区、康复区、护理与后勤区、中心花园区。各功能区布局合理，联系便利，互不干扰。

现已建成门诊楼和住院楼。门诊楼为综合性大楼，共11层，集综合门诊、专科门诊、急诊、医技、药房、办公、学术交流等功能为一体，各流线相互分开，互不干扰。内部设有不同尺度的中庭，丰富内部空间，也满足内部自然采光需求。

住院楼共9层，设有住院病房、中心供应、ICU、手术室等功能，共设有11个护理单元。各区域设有轨道物流传输系统，医疗用品可从中心供应到达各功能区域，方便快捷，提高效率。

Wuhu Conch Hospital, a 1,500-bed general hospital in Wuhu, Anhui Province, is divided into five functional zones in master plan, namely, outpatient and medical technology, inpatient, rehabs, nursing & BOH, and central garden based on careful consideration to the topography, context and road traffic conditions. The functional zones are distributed quite reasonably to facilitate interconnection and avoid mutual interference.

So far, the outpatient and inpatient buildings have been completed. The 11-floor outpatient building integrates functions like the general outpatient, specialist outpatient, emergency, technology-aided diagnosis, pharmacy, offices, academic exchanges, and so on. Circulations of these functions are separated from each other to avoid interference. Besides, atriums in varied sizes in the building not only diversify the interior spaces but also let in more daylighting.

The 9-floor inpatient building houses inpatient wards, central supply, ICU, operating room, and 11 nursing units. All the units are equipped with track logistics system, by which medical necessities in central supply can be conveniently and efficiently sent to different zones.

项目地点：安徽省芜湖市
设计时间：2012
建设时间：2016.3
建筑面积：198000m²
建筑层数：地下1层，地上11层
建筑高度：51.3m

Location: Wuhu, Anhui Province
Design: 2012
Construction: 2016.3
GFA: 198,000m²
Floors: 1 underground, 11 aboveground
Height: 51.3m

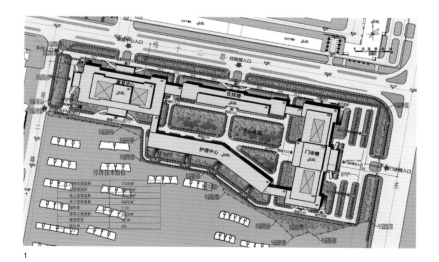

1

2

1 总平面图
  Site plan

2 门诊主入口突出，外立面采用GRC材料，统一协调
  The outpatient building with a distinct main entrance features GRC façade for a consistent and harmonious appearance

3 医院交通便利，人流和车流分开设置
  The hospital enjoys convenient transportation with separate pedestrian and vehicular circulation

4 由住院楼和门诊楼围合而成的内庭院，环境优美
  Inner courtyard surrounded by inpatient and outpatient buildings offers attractive environment

地下空间

2005年GDAD参加广州珠江新城花城广场（珠江新城核心区市政交通项目）国际招投标，并最终获得设计权，历时五年的设计和建设周期，为广州市打造了全国著名的新城市名片，提升了广州城市环境的核心竞争力。借此机会，GDAD由院领导亲自挂帅组成了一个集城市规划、市政、道路、建筑设计、智慧城市、市政管廊与一体的地下空间研究和设计团队，为广州市及国内部分城市提供地下空间的技术服务，并在理论研究、专利申请、论文发表和专项研究方面取得了广泛和突出的成果。

近年来，GDAD通过国际和国内招投标，承担了包括广州珠江新城花城广场、番禺万博商务中心、时尚天河一期工程、广州国际金融城起步区、东莞南城商务区、广州琶洲电商服务区西区、广州南沙明珠湾起步区、广州江南西路、成都光华新区、惠州市民广场、广州岭南广场等近300万平方米的多项城市核心区重大地下空间项目的设计任务，参加了广州火车南站、广州白云山北门、南京江北新区起步区、宁波新世界等地下空间项目的前期研究工作，在城市地下资源整合、城市地下空间开发时序、城市地下空间建设模式、城市地下空间技术措施、城市地下空间绿色环保、城市地下空间防灾防涝、城市地下空间投资控制等方面形成了一整套的理论框架和研究，并经受了实践的检验。

在积极参加城市地下空间建设实践过程中，GDAD作为主要参编单位，编制了行业标准《城市地下空间利用基本术语标准》及国标《地下建筑设计统一规范》，主编了《地下空间有序建设配套政策研究》、《珠江新城核心区市政交通项目设计指引》、广州国际金融城起步区地下空间9大导则指引及技术措施《技术导则》、《设计指引》、《防水技术构造措施》、《消防设计技术措施》、《地下市政道路设计技术措施》、《抗裂减振专项技术措施》、《垂直交通技术措施》、《建筑通用构造大样选型图集》、《建筑材料应用与设备选型规定》等技术研究文件，并成功申请五项国家专利、获得多项国家及省级科技进步奖。

地下空间开发作为节约土地资源、改善城市交通环境、城市资源综合利用的重要举措，在我国城市化建设的进程中越来越受到广泛重视，GDAD作为国内在此领域较早具有理论和实践相结合的全专业综合设计研发机构，将一如既往的在未来城市地下空间建设方面，为社会各界提供优质、创新、专业的技术服务，打造一个又一个环境优异、功能合理、交通便捷、效益兼顾、安全可靠的地下城市。

Underground Spaces

In 2005, we participated in the international competition for the Huacheng Square, Zhujiang New Town, Guangzhou (Municipal Transportation Project in Core Area of Zhujiang New Town, Guangzhou), and was finally awarded the contract. After a five-year design and construction period, the Project was completed as a new icon of the city to enhance the core competitiveness of the city's urban environment. Taking this opportunity, we established an underground space research and design team under the direction of its top management. This multidisciplinary team includes urban planning, municipal utilities, roads, architecture, smart city and utility tunnel, aiming to provide technical services of underground spaces for Guangzhou and some other cities in China. So far the team has achieved extensive and remarkable results in terms of theoretical research, patent application, paper publication, and special research.

In recent years, we have, through international and domestic design competitions and bidding, undertaken the designs for various underground spaces in the core urban areas with a total of nearly 3 million square meters, including
-Huacheng Square, Zhujiang New Town, Guangzhou;
-Wanbo Business Center, Panyu District;
-Phase I of Fashion Tianhe Plaza;
-Kick-off Zone of Guangzhou International Financial City;
-South City Business District, Dongguan;
-West Area of Pazhou E-commerce Business Area, Guangzhou;
-Kick-off Zone of Pearl Bay, Nansha District, Guangzhou;
-Project of Jiang Nan Xi Lu, Guangzhou;
-Guanghua New Area, Chengdu;
-Civic Square, Huizhou; and
-Lingnan Square, Guangzhou.

We also participated in the early research work of such underground spaces as
-South Railway Station, Guangzhou;
-North gate of Baiyun Mt, Guangzhou;
-Kick-off Zone of Jiangbei New Area, Nanjing; and
-New World, Ningbo,

So far we have developed a full set of theoretical framework and research results which are proven workable in term of integration of urban underground resources, development sequence, construction mode, technical measures, environmental protection, disaster and waterlogging prevention, and investment control of urban underground spaces.

While proactively participating in the construction of urban underground spaces, we co-edited *Standards for Basic Terminologies of Urban Underground Spaces*, an industrial standard, and *Unified Code for Design of Underground Buildings*, a national standard, and compiled various technical research documents, including *Study of Supporting Policies for Orderly Construction of Underground Spaces, Design Guidelines for Municipal Traffic in Core Area of Zhujiang New Town*, and nine major guidelines and technical measures for underground spaces of the Kick-off Zone of Guangzhou International Financial City, namely, *Technical Guidelines, Design Guide, Waterproof Technical Measures, Technical Measures for Fire Protection Design, Technical Measures for Design of Underground Municipal Roads, Specialized Technical Measures for Crack Resistance and Vibration Reduction, Technical Measures for Vertical Transportation, Atlas of Generic Building Construction Details, and Regulations on Building Materials Application and Equipment Selections*. In addition, we have registered five national patents and won various technological progress awards of national and provincial level.

As an important measure to save land, improve urban transportation and properly utilize urban resources, development of underground spaces has been increasingly emphasized in China's urban development. We will take full advantage of our leading position as an multi-disciplinary design and research institution with years of theoretical and practical experiences in this field, and, as always, provide quality, innovative and specialized professional services for the future development of urban underground spaces and create attractive, functional, accessible, economical, safe and reliable underground urban spaces.

广州市珠江新城核心区市政交通项目
Municipal Transportation Project in Core Area of Zhujiang New Town, Guangzhou

广州市珠江新城核心区市政交通项目是广州市政府为配合2010年亚运会召开的重点工程之一，是广州市目前已建成规模最大、最重要的地下空间开发。项目地下总建筑面积约50万平方米，通过三层地下空间达到疏导交通与商业发展双赢的有效结合，更将周边共120万平方米、39栋商业办公楼、公共文化建筑的地下空间资源连通整合。

项目通过下沉景观广场沿中轴线布局，构成地下商业城的脊柱。下沉景观广场、大型坡道和楼梯，将自然景观绿化引入地下，使地下空间与地面建筑和景观从视觉和空间上融合为一体。地下的商业购物廊犹如血管延伸在地下各个功能区，围绕着联系轨道交通及周边建筑地下空间的人行通道系统展开，使地下人行系统在空间和装饰上产生人性化建筑效果。纵横交错的购物街，将地下空间内不同的功能区域有机的连接起来，形成一个连续的、有趣的、具有动感的建筑和景观空间。

珠江新城核心区市政交通项目连接、整合区域内各类综合设施和周边建筑的地下空间，统一规划区域内供电、给排水、供冷、垃圾收集、安全监控系统、消防设施、人防设施、停车库等各项设施，使区域内各项公共设施达到统筹、统一、有机结合。形成一个资源共享的地下公共空间体系。

Municipal Transportation Project in the Core Area of Zhujiang New Town, Guangzhou is one of the key project of the city government supporting 2010 Asian Games and the largest and most important underground space development completed in Guangzhou so far. The underground gross floor area of the project is 500,000m² and three underground floors are developed to effectively cater to two purposes: transportation and commercial development. It also connects and integrates the underground spaces of 1.2 million m² under 39 office buildings and public cultural buildings in the district.

The project stretches along the central axis via sunken landscaping plaza to create the backbone of the underground shopping district. The sunken landscaping square, generous ramps and staircases bring natural landscape and greening into underground space and help visually and spatially integrate below-grade space with above-grade architecture and landscape. Underground shopping arcades extend along the pedestrian passages system that connects to metro station and the underground spaces of surrounding buildings, contributing to the people-oriented architectural effect of the pedestrian system in terms of space and finishing. The crisscrossed shopping arcades bring various functional zones in underground space together, shaping a continuous, engaging and dynamic architectural and landscape space.

The project integrates and connects utilities and underground spaces of surrounding buildings in the district, meanwhile synergizes the planning of the electricity, water supply and drainage, cooling, waste collection, fire protection, civil defense, parking and other facilities to create an integrated, uniform and interconnected utility system and a resource-sharing underground public space system.

项目地点：广州市珠江新城
设计时间：2006.5–2007.10
建设时间：2012.11
建筑面积：370000m²
建筑层数：地下3层，地面1层
建筑高度：5.5m
合作单位：欧博迈亚设计咨询有限公司
曾获奖项：2013年度广东省优秀工程勘察设计奖公建类一等奖
中国建筑设计研究院CADG杯华夏建设科学技术奖三等奖
第七届全国优秀建筑结构设计奖二等奖

Location: Zhujiang New Town, Guangzhou
Design: 2006.5–2007.10
Construction: 2012.11
GFA: 370,000m²
Floors: 3 underground, 1 aboveground
Height: 5.5m
Partner: OBERMEYER
Awards: The First Prize of Public Building under Excellent Engineering Design Award of Guangdong Province, 2013
The Third Prize of CADG Cup - Award for Science and Technology Advancement;
The Second Prize of the Seventh National Excellent Building Structure Design Award

1 总平面图
  Site plan

2 广州塔上鸟瞰整个项目，可瞭望城市中轴线上城市的空间序列
  A bird's eye view of the whole project from Canton Tower; a view of the urban spatial sequence along the city's central axis

3 作为广州市新中轴上标志性"城市客厅"的夜景
  Night view of the representative "city showcase" on the new city axis

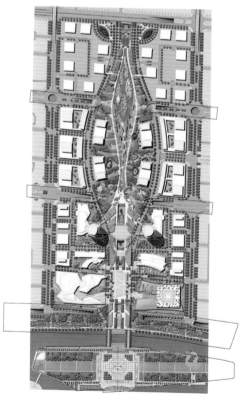

1

金融城地下空间
Underground Space of Guangzhou International Financial City

广州金融城起步区公共地下空间项目是实行三维立体式开发、功能复合化的地下综合体，代表当今世界最先进地下空间开发水平，拥有最完善的地下空间功能。拟打造融合交通、商业、公共服务、停车、绿色市政的复合型地下城市综合体，形成配套完善、功能复合、上下一体、交通便利、安全舒适的地下城市网络。

项目根据规划形成"三核三轴七组团"的地下空间开发结构。以道路交通线路为骨架，形成枢纽核心、翠岛核心、方城核心及七个组团空间，核心区通过商业步行街形成的地下发展轴与其他组团相联系，形成共同发展的地下空间网络。坚持集约建设原则，发挥交通枢纽、地铁站的辐射价值，实现地下空间公共地块与开发地块功能、空间的无缝衔接。优化提升区域交通，形成集地铁、公交、车行、人行的一体化立体综合交通网络。充分利用现状河涌、结合广州气候特点，塑造下沉广场、步行商业街，显现岭南开放兼容的核心特质。加强人工环境与自然环境的交互，加强地下景观塑造，创造生态、舒适的地下环境。

金融城起步区公共地下空间旨在形成一个活力、高效、低碳示范的地下枢纽，通过土地交通、资源利用、环境模拟、绿色建筑四个低碳生态角度的研究，将新技术、新理念落实到土地空间开发、交通、市政等专项规划，统筹地下市政设施，采用成熟的绿色市政技术集成，形成低碳环保、高效节能、资源共享的示范区。为金融城营造出回归自然、回归生活的理想空间，并作为绿色生态示范城区与绿色建筑示范区凸显出广州新型城市化的最新成果。

The underground space project in Kick-off Zone of Guangzhou International Financial City is a mixed-use multi-level underground complex. It represents the world's most up-to-date underground space development and offers the full-fledged underground spatial functions. As a multi-purpose complex below grade, it integrates transportation, retail, public services, parking and greening utilities, offering a safe and comfortable underground urban network that features well-established facilities, multifold functions, synergy with above-grade developments and convenient transportation.

The project is planned into a pattern of Three Cores, Three Axes and Seven Clusters. With the road network as backbone, the central core, Cuidao core area, Fangcheng core area and seven clusters are created. The core areas are connected to other clusters via the underground axes of the pedestrian shopping arcades, establishing the interconnected underground space network. Following the principle of intensive development, the project makes full play of transportation hub and metro station to ensure seamless integration of functions and space between the public underground space and site. By optimizing and improving regional transportation facilities, an integrated, multi-level and multi-modal transportation network is forged to integrate the metro, buses, private cars and pedestrian circulations. Existing canals and local climate conditions are fully considered in design of the sunken plaza and pedestrian shopping arcade to highlight the openness and inclusiveness of Lingnan culture. The alternating artificial and natural environment contribute to shaping the underground landscape and an eco-friendly and comfortable underground environment.

The project aims to create a vibrant, efficient and low-carbon underground hub. With researches on low-carbon ecology in four areas, namely land resources and transportation, resources utilization, environmental simulation and green building, the project implements the new technologies and concepts in specialized planning for land and space development, transportation and utilities etc. Meanwhile, underground utilities are provided as a whole and mature green technologies adopted to set up an example for low-carbon, environmental-friendly, energy efficient and resource-sharing development. While creating an ideal space in the Financial City, the project is anticipated to be a retreat from the bustle and hustle of the city, an example of green urban area and green building, and the city's latest result in new urbanization.

项目地点：广州市金融城
设计时间：2014
建设时间：2015
建筑面积：440000m²
建筑层数：地下4层，地面1层
建筑高度：5.5m

Location: Guangzhou International Financial City
Design: 2014
Construction: 2015
GFA: 440,000m²
Floors: 4 underground, 1 aboveground
Height: 5.5m

1 方城广场剖面图，高效畅通的地下道路网络实现地面步行空间的完整
  Sectional view of Fangcheng Plaza: highly efficient and smooth underground road network ensures the integrity of above-grade pedestrian space

2 花城大道剖面效果图，开放式的地下公交站场与地下轨道交通"零"换乘
  Sectional view of Hua Cheng Da Dao: open underground bus depot and underground metro station ensure easy transit

3 方城广场地下三层公交站场空间效果图，方城下沉广场作为金融城中央地下交通网络的中心枢纽
  Rendering of bus depot on B3 of Fangcheng Plaza: Fangcheng Sunken Plaza serves as the hub of the underground transportation network of the Financial City

4 方城广场地下二层架空空间效果图
  Rendering of open-up space on B2 of Fangcheng Plaza

5 花城大道绿融路路口剖面图，整合地下公交、人行、地铁、隧道以及综合管沟功能
  Sectional view of the intersection of Hua Cheng Da Dao and Lyu Rong Lu: The underground bus, pedestrian, metro, tunnels and multi-functional piping gallery are integrated

1

2

3

4

5

万博地下空间
Underground Space of Wanbo Business District

万博商务区地下空间项目属于广州首个地面地下统筹设计、同步开发实施的地下空间综合开发项目。利用公共地下空间、整合、连接核心区地下空间建筑及交通、市政、综合防灾各设施，形成一个统一、有机的公共空间体系。

设计理念定位于打造一座"大隐于市的地下之城"。作为核心区地下建筑相互联系的纽带，为各区域地下空间搭建有机骨架，形成地下之城连贯而自成系统的主要脉络。项目由十字主轴、地下主环、公共交通节点等组成，形成了"十字主轴、交通管线环路"的结构。室内人行空间采用自然、舒适的地下空间设计，以退台方式构造可自然采光的半地下空间，突出岭南建筑风格，形成独特的特色地下骑楼街；利用道路的中央绿化带设置地下绿化中庭，实现地下步行通道的自然通风采光。设计合理组织了立体的地下综合交通，设法降低地面交通的拥挤程度，使项目用地的公交和步行交通出行率达到最大化。其中，地下一、二层为人行通道及配套商业，地下三层设置地下环路及停车，局部地下四层为地铁平台层及停车。

万博地下空间项目的建设将增强万博中央商务区的交通功能，加强与外围城市交通的衔接和联系，实现与公交、轨道交通的便捷换乘功能，创造多层次的地下立体交通体系；连接、整合区域内各类综合设施和周边建筑的地下空间，统一规划区域内供电、给排水、供冷、垃圾收集、消防设施、人防设施、停车库等各项设施，使区域内各项公共设施达到统筹、统一、有机结合，形成一个资源共享的地下公共空间体系。

The underground space of Wanbo Business District is the first underground development project in Guangzhou that is designed and developed in synergetic and synchronized manner with above ground projects. Public underground spaces are utilized and integrated to connect buildings in the core area and to provide transportation, municipal and comprehensive disaster prevention facilities, hence forming a synergetic and organic public space system.

The concept of the design is to build an invisible underground city. As a link that connects major buildings in the core area, the project aims at building an organic framework for underground spaces of various areas and shaping the skeleton of a well-connected and systematic underground city. The project features a structure of crossed axes plus ring roads and pipelines, consisting of two crossed axes, main underground ring road and public transportation nodes. Indoor pedestrian space is designed to ensure the daylight and comfort. The daylit semi-underground spaces feature the recessed ground floor, which represents a typical Lingnan style and gives the unique impression of an underground arcade street. Central green belts along roads are designed into courtyards to bring daylight to underground pedestrian passages. Well-organized three-dimensional underground transportation facilities reduce congestion above ground and maximize the efficiency of public transportation and pedestrian access of the site. B1 and B2 are designed into pedestrian passages and commercial facilities, B3 provides car parks and underground ring roads while partially developed B4 is designed into metro stations and car parks.

The implementation of Wanbo Underground Space project is expected to enhance the transportation convenience of Wanbo Central Business District, strengthen its links and connections with surrounding urban areas, ensure efficient transits between buses and rail transportation and fulfill the target of building a multi-level underground transportation system. By integrating and connecting various comprehensive facilities and underground spaces of buildings in the region, and thanks to the synergetic planning of power supply, water supply and drainage, cooling, waste collection, fire protection, civil defense facilities, parking and other facilities etc, an integrated, uniform and organic public service system is ensured and a resource-sharing underground public space system is formed.

1

2

项目地点：广州市番禺区
设计时间：2013
建设时间：2014
建筑面积：390000m²
建筑层数：地下4层，地面1层
建筑高度：5.5m

Location: Panyu District, Guangzhou
Design: 2013
Construction: 2014
GFA: 390,000m²
Floors: 4 underground, 1 aboveground,
Height: 5.5m

1 总体鸟瞰图，地下空间整合周边九大地块的地下资源
 A bird's view of the site; the project integrates underground spaces of nine major plots in the region

2 总平面图
 Site plan

3 地下空间剖面图，结合周边项目地下空间的开发，形成万博商务区"大隐于市的地下之城"
 Sectional view of the underground space; with the development of underground space of surrounding projects, together they form an invisible underground city under Wanbo Business District

居住建筑

从巢穴而居、结草为庐，到倚木干阑、豪门宅第，再如今砼石成林、高楼梦幻。随着社会不断发展，人们对归属感、安全感、幸福感有了更丰富的内涵需求和期望。GDAD的住区规划、居住建筑设计也不断迭代更新，相应的设计理念亦在与时俱进。GDAD努力把握住区开发与城市发展的互动关系，以及由此所形成的供需关系规律，在尊重自然、尊重文脉的原则上，主动创造居住建筑的产品形式，用现代语汇设计宜居的高尚住区。设计作品中，灵活运用自然生态美学与建筑类型学的理念，用物质的实体来表达抽象的诗情画意，又绝不刻板地复制传统营造手法，力求感受天地大象大美，洞悉大自然的本质规律。

在设计作品中，附加产品特性的住区规划、居住建筑设计呈现多样化的趋势，逐步走向模式化、精细化、主题化。从"健康住宅"、"绿色住宅"、"生态住宅"到"亲情住宅"、"第二居所"等等，新概念层出不穷。从美伦公寓的山水庭园，到依云水岸的空中花园，再到伍兹公寓的顶层泳池，在注重对居住的舒适性、健康性的设计追求同时，对居住建筑产品的创新性越发关注。

《周礼》："天生时，地生气，材有关，工有巧，合此四者以为良。"在技术设计过程中，GDAD结合对装配式建筑技术的研发应用，探索居住建筑建设工业化模式与传统营建之间的关系；结合BIM对居住建筑精细化设计，逻辑推演建造过程与完成形式之间的完美关系。

居住建筑不仅提供"安身之所"的基本居住需求，而且是文化的重要载体，是所处时代生活文化的物化。GDAD对传统居住文化追本溯源，运用"源于自然"自然生态美学观，积极探索传统人居需求的现代语境表达，将中国传统审美观和现代生活价值观进行高品质融合，提炼输出具备典型代表性的规划理念及产品方案，力求构建结合现代场景需求、满足现代心灵回归传统的理想住所。

如鲸山九期花园的"原石"概念是对独特的岩石景观的自然生态美学诠释；方框形构件模块组合、叠合，表述一种"自然生长"的语境。方框形构件的"方"构形，源自于岭南传统民居三间两廊的"方院"，透露出"家园"的隐喻，也成为住区场所氛围构成的核心词汇。

在设计作品中，多从类型学的思考角度抽象共性特点，运用邻里单元空间组合实现多样性的聚落型形态；注重对住区环境自然特征的提取和营造，思考回归绿水青山和泥土气息的生态自然，回归本土和地域的文化本源。

如美灵湖住区湖光山色、金山谷花园鸟鸣翠谷。如在猎德旧改中平面脉络"龙舟竞发"，提炼岭南元素，将传统空间与现代居住元素进行延续、抽象与提炼，形成具备历史底蕴的、更适宜居住的新中式风格的生活范本。如鲸山九期花园的建筑布局依山、观海、瞰城，"居于半山之上，层楼掩映于林"是项目区别于其他楼盘的特色，也是项目策划之初就希望能够传达给客户的社区场景。住区居住生活与自然生态系统、地域文化底蕴融为一体，人们不仅有亲近自然的乐趣，还增加自身文化的标识性与自豪感，体现中国住区环境、居住建筑营造的"人居相依"境界。

Residential Buildings

From nesting in caves to dwelling in thatched cottages, houses perched on stilts to luxury mansions, and bristly concrete buildings to fancy skyscrapers, the constant development of society has brought richer meaning and higher expectation to people's sense of belonging, security and happiness. We are also iteratively updating our residential district planning and residential building design and keep our design concepts abreast of the times. Focusing on the interactive relations between residential district development and urban development, and taking into consideration the corresponding laws of demand and supply, we have created residential building products with respect for nature and cultural context to provide livable high-end residential building design through modern architectural vocabulary. In the design, we employ the concepts of ecological esthetics and architectural typology to express the abstract poetic beauty through physical objects, while avoiding a mere mechanical copy of the traditional construction techniques, in an effort to appreciate the profound image and great beauty of heaven and earth and discover the intrinsic laws of nature.

In our design works, residential district planning and residential building design with product properties are taking on a trend of diversification and becoming more modular, refined and theme-focused. From healthy housing, green housing, ecological housing to family housing and a second home, new concepts have been emerging one after another. From the courtyard with mountain and water views in Maillen Apartment to sky gardens in Evian Town, and top floor swimming pool in Woods Park, more attention has been paid to the innovation of the residential building, in addition to comfort and healthy living.

The Rites of Zhou says that *time, place, selection of material and workmanship jointly make excellence*. In our technical design process, we explore the relations between the industrialization of residential building development and the traditional construction techniques based on research, development and application of prefabricated building technology, and logically extrapolate the perfect relation between the construction process and the completed form through the BIM-based delicacy design of residential buildings.

Residential buildings are not only a place of shelter to meet the basic need of living. They also function as a carrier of culture, a product of life and culture of a specific time. Tracing back to the sources of traditional residential culture and employing the natural and ecological aesthetic concept of *natural origin,* we actively explores the expression of traditional residential demands in a modern context, properly integrate the traditional aesthetics of China with values of modern life, and extract and present representative planning concepts and product proposals , striving to create ideal residences that meet the contemporary demands and people's desire to return to tradition.

The concept of natural rock in Jingshan Garden Phase IX, for example, is an aesthetical interpretation of the unique rockscape in natural and ecological form, where the box-shaped component modules are assembled and stacked to present a setting of natural growth. In fact, the square shape of the component is originated from the quadrangle with three houses and two corridors in traditional Lingnan houses. It is a metaphor for home and also a key word for the ambience in the residential district.

For most of our design works, we abstract the common properties from the perspective of typology, and realize the diverse settlement through configuration of various neighborhood spaces; we emphasize the extraction and creation of natural environment in residential districts, with the purpose of returning to the green mountains, clear waters and fresh earth of nature and regressing to local and regional culture.

In the beautiful landscape of lakes and mountains in Meiling Lake Residential District, the green valleys and chirping birds in The Hills, planar composition symbolizing the competing dragon boats in the renovation of Liede Village, the Lingnan elements are extracted and traditional spaces and modern residential elements are extended, abstracted and extracted into an example of a more livable Neo-Chinese-style residence with rich cultural implications. The Jingshan Garden Phase IX, for example, nests in the mountain, facing the sea and overlooking the city. *Living halfway up the mountain in buildings nestling amidst the trees* is what make it distinguished from other projects, and it is also a scene of the community expected to be delivered to clients at the initial design phase. Life in the residential district is closely integrated with nature, ecology, and regional cultural implications, so that people can not only embrace the nature, but also share an increased sense of cultural identity and pride. This reflects a state of interdependency between people and residence within the human settlement and residential buildings in China.

广州猎德村旧村改造项目
Reconstruction of Liede Village, Guangzhou

猎德村位于广州CBD(中央商务区)的核心，建村至今约九百年的历史，成为广州城市历史悠久、文化底蕴丰厚和生生不息的象征。猎德村改造复建完成后有利于完善珠江新城的交通体系，改善中央商务区的面貌，培养现代服务业氛围。作为一个城中村改造项目，桥东安置区是猎德旧村改造复建项目的核心。对其改善村民生活条件、均衡补偿及分配原则、完善城市公共体系、传承地域文化等具有积极意义。

猎德村改造复建分三部分进行规划：桥西区(用地约11.4万平方米)按价值最大化原则进行拍卖；用于发展商务的桥西南区(用地约5万平方米)规划五星级酒店作为村集体经济的支撑项目；桥东区做为村民的安置区(用地约17.1万平方米)。项目充分利用地块的自身价值，综合开发，完成具有社会和经济双重效益的综合项目。桥东安置区利用桥西区拍卖的资金进行新猎德村的复建，新村设计总人口约2万人，建筑总面积93.4万平方米。整体复建改造工作于2010年基本完成，并于同年9月28日入住。桥东安置区的总规划构思提炼旧村的网格状的城市肌理，以复建的宗祠区作为传统文化的承载，猎德涌和复建的池塘作为龙舟文化的源头，形成犹如龙舟待发的六行建筑群，组成"龙舟竞发"的总平面脉络，民俗文化传承渗透其间，表现了城市化进程中的现代猎德村的整体性及一体化； 岭南建筑元素的提炼与现代建筑元素的结合手法融合到安置区的规划，宗祠区、 文化长廊、 龙舟湖、 猎德涌、 牌坊、 建筑设计、建筑造型创作及园林设计中，充分体现尊重村的传统文化及生活模式的理念， 孕育出新的现代猎德村。

猎德村对广州城市发展有着深远的影响，它不但是一个旧村改造工程，也是一个具有深刻社会意义的工程。

Liede Village is located at the heart of Guangzhou's CBD. The village has a history of about nine hundred years, which is long enough to represent Guangzhou's profound history, extensive culture and everlasting vitality. The reconstruction of Liede Village is conducive to improving the transport system in Zhujiang New Town, and thereby upgrading the appearance of the CBD and creating an amicable environment for modern service industries. East Bridge Area for resettlement is the core of the Project, which can significantly better the living conditions of villagers, ensure balanced compensation and distribution, optimize the urban public system, and carry forward the regional culture.

The reconstruction of the Liede Village was planned into three parts: West Bridge Area (about 114,000m²) was to be auctioned by maximum of value; Southwest Bridge Area (about 50,000m²) was designed with a five-star hotel that would underpin the collective economy of the village; East Bridge Area (about 171,000m²) was to accommodate the resettled villagers. By fully exploiting the value of the land through comprehensive development, the Project achieved both social and economic benefits. Using the funds raised from the auction of the West Bridge Area, East Bridge Area was reconstructed to accommodate 20,000 or so resettled villagers on a total floor area of 934,000m². The overall reconstruction was completed in 2010 and the villagers moved in on September 28 in the same year. The East Bridge Area was planned based on the grid-like urban fabric of the old village. With the reconstructed ancestral hall area serving as a carrier of the traditional culture and Liede Canal and the reconstructed ponds as the origin of the dragon boat culture, six rows of building clusters resembling some competing dragon boats are created in the master plan. Folk culture is also incorporated to showcase the integrity and integration of modern Liede Village in the process of urbanization. In addition, refined elements of Lingnan architecture and modern architecture elements were mixed together in the planning of the resettlement area and the design of the ancestral hall area, the culture corridor, the dragon boat lake, Liede Canal, the memorial gateway, architectural form and landscape, fully embodying the respect for the village's traditional culture and lifestyle and creating a new modern Liede Village.

The Project has far-reaching impacts on Guangzhou's urban development initiative. It is not simply about urban village reconstruction, but more importantly a project of profound social significance.

项目地点：广东省广州市天河区珠江新城
设计时间：2007- 2008
建设时间：2008- 2010.10
占地面积：133000m²
建筑面积：93.4万m²
建筑高度：6- 120m
建筑层数：1- 39层（包括复建宗祠区）
曾获奖项：2011年全国优秀城乡规划设计（村镇规划类）三等奖
　　　　　2011年广东省优秀工程勘察设计一等奖
　　　　　2011年广东省城乡规划设计优秀项目二等奖

Location: Zhujiang New Town, Tianhe District, Guangzhou, Guangdong Province
Design: 2007-2008
Construction: 2008-2010.10
Site: 133,000m²
GFA: 934,000m²
Height: 6-120m
Floors: 1-39 (including the restored ancestral hall area)
Awards: The Third Prize of National Excellent Urban and Rural Planning Award (Village Planning), 2011
The First Prize of Excellent Engineering Exploration and Design Award of Guangdong Province, 2011
The Second Prize of Excellent Urban and Rural Planning Projects of Guangdong Province, 2011

1　总平面图
　　Site plan

2　自猎德东西向牌坊望新村视点
　　Views from the east and west of Liede Memorial Gateway to the new village

3　复建的建筑宗祠群，夕阳下岭南古建筑和现代简约建筑相得益彰
　　The reconstructed ancestral halls of ancient Lingnan architectural style setting off modern concise buildings in the sunset

4　沿珠江展开的安置区住宅，每组建筑之间为视觉和通风通廊，在高容积率的要求下尽量减少对城市对景观影响
　　Resettlement area expanding along the Pearl River, with corridors for vision and ventilation from building to building to minimize impacts on the urban landscape while maintaining high FAR

5-7　自珠江边往猎德大牌坊视点，建筑群高低错落有致
　　　3. Views of the well-arranged buildings from the Pearl River bank to Liede Memorial Gateway

1

2

3

4

5

6

7

广州招商金山谷花园(1-4期)
The Hills (Phase I- IV), Guangzhou

金山谷花园倡导绿色生态社区规划及建筑设计，项目依照可持续发展的开发理念，以循环经济、新都市主义、理性增长为指导原则，引入生态足迹、"一个地球生活"即O.P.L理念的十项原则、可持续发展设计等先进开发技术，实现生态最大化、社区综合开发，并成为低碳社区。

项目的总体布局由不同住宅类型结合自然地形进行小区设计，因地制宜，尊重原生态，以用地的两座山为核心，形成一片绿意盎然、分别有高、低层的住宅组团，依山而建的山地社区。小区规划尽量保留山顶的原始植被，形成原生态的山顶森林，同时保护用地中有价值的树木，形成林中小屋和树下人家的写意生活。O.P.L理念的十项原则：零碳排放、零废弃物排放、可持续发展交通、可持续发展和本地材料、本地食物、节水、保护生物资源、传承文化、公平贸易、分享健康愉快的生活。同时，金山谷项目自2005年开展的生态规划到现时1-4期已投入使用，已用及拟用的绿色技术包括20项，包括：生态最大化设计、计算机模拟辅助设计、中水回用、风力发电技术、自然采光、绿色照明、屋顶绿化、高热工性能围护结构、室外热环境改善、节水器具、太阳能光热利用、室内空气品质、室内通风模拟、室内空气品质监测系统、住宅能耗标识、环保建材—再生塑木地板、可调节外遮阳措施、旧建筑利用、节能设备—全热新风交换机、高效供水设备等。

让人远离大都市的喧嚣和生活的快节奏，享受恬静、温馨和悠闲的社区生活和回归自然。在城市化进程愈演愈烈的今天，规划建筑设计在担负起修复被破坏的人类居住环境和慰藉希望复旧自然的心灵的使命。借此项目，使金山谷达到"网罗天地于门户，饮吸山川于胸怀"的回归自然的高尚居住境界。

项目地点：广东省广州市番禺区
设计时间：2008.2-2009.12
建设时间：2009-2012
建筑面积：667000m²
建筑高度：12-100m
建筑层数：3-32层
曾获奖项：2009年联合国人居署颁发的"可持续人居最佳范例奖"（UN HABITAT-A BEST PRACTICE UN-HABITAT BISINESS AWARD)（中国第一次获此国际大奖）
2010年全国人居经典建筑规划设计方案竞赛综合大奖
2010年最佳低碳社区奖
2013年度广东省优秀工程勘察设计奖工程设计二等奖
2013年度全国优秀工程勘察设计行业奖住宅与住宅小区三等奖

Location: Panyu District, Guangzhou, Guangdong Province
Design: 2008.2-2009.12
Construction: 2009-2012
GFA: 667,000m²
Height: 12-100m
Floors: 3-32
Awards: UN HABITAT- A Best Practice UN-HABITAT Business Award) (the first of its kind in China)
Best Overall Award of National Planning and Design Competition for Classic Residential Buildings, 2010
Best Practice for Low-Carbon Society, 2010
The Second Prize of Excellent Engineering Exploration and Design of Guangdong Province – Engineering Design, 2013
The Third Prize for Residential Buildings and Quarters of National Excellent Engineering Investigation and Design Industry Award, 2013

To implement eco-community planning and green building design, The Hills Project, under the concept of sustainable development and the guiding principles of circular economy, new urbanism and rational growth, has brought in many advanced development technologies including ecological footprint, the ten principles of "One Planet Living", i.e., the O.P.L concept, sustainable development design etc. for the purpose of realizing maximum ecological preservation, overall community development and a low-carbon community.

The general layout of the community is so designed that various types of residential buildings are delicately integrated into existing natural landscape with minimum impact on original ecology. With two hills as the center, a hillside community in exuberant greenery with residential complex of both high-rises and low-rises is established. Existing vegetation on top of the hill is preserved to the maximum in design to serve as a natural mountain forest, and valuable trees on site are also retained to create an idyllic picture of cabin in the woods and life under the trees. The ten principles of O.P.L concept are zero carbon, zero waste, sustainable transport, sustainable materials, local and sustainable food, sustainable water, land use and wildlife, culture and community, equity and local economy, and health and happiness. Meanwhile, Phase I-IV eco-planning of the Hills initiated in 2005 has been put into operation, with 20 green technologies applied or to be applied, including maximum ecological preservation design, computer simulation aided design, reuse of reclaimed water, wind power technology, daylighting, green lighting, roof greening, high thermal performance envelop, external thermal environment improvement, water-saving facilities, solar light and heat utilization, interior air quality, interior ventilation simulation, interior air quality monitoring system, residential building energy-efficiency label, green building materials-recycled wood plastic flooring, adjustable external sunshades, utilization of existing buildings, and energy-efficient equipment such as air to air total heat exchanger and high efficiency water supply facilities etc.

As a retreat from the city's hustles and bustles, the project offers the residents a tranquil, cozy and leisure community life. In a time of unprecedented urbanization, the architectural design and planning will contribute to restoring the damaged living environment and comforting the souls longing for nature. The project of the Hills sets up an example of natural residential environment where the nature is just within an arm's reach.

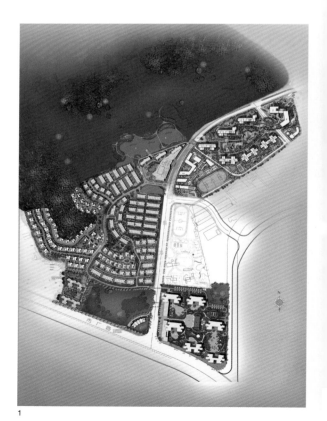

1

1	总平面图 Site plan
2	金山谷四期富有格调的小区主入口 Stylish main entrance to The Hills Phase IV
3	"生态休闲"联排别墅 "Eco-leisure" townhouse
4	情景别墅围合园林空间 Villa-framed garden space
5	全新南坡地别墅组团"楹"及"尊贵"豪华高层 New hillside villa cluster named Ying and luxury high-rise
6	二期跃式高层住宅 Phase 2 skip-floor residential high-rise

深圳招商美伦公寓
Maillen Apartment, Shenzhen

美伦公寓项目坐落于深圳市蛇口半山区，地块东侧为沿山路，西面为半山海景别墅，北面是半山社区中心和兰溪谷高档住宅区。蛇口位于深圳南头半岛东南部，东临深圳湾，西依珠江口，与香港西部隔海相望，是我国对外改革开放的前沿阵地。经过三十多年对外改革开放的经济建设，蛇口已发展为孕育近百家国际著名企业的国际化社区，是深圳与国际接轨最近的地方，美伦公寓项目定位为面向海外高级人才的高档精品小户型出租公寓。

美伦公寓设计构思希望在城市整体形象由于全球化和商业化所驱使而变得"一般化"、"相同化"的情景下，通过创造一种居住模式来反映地域化、个性化。设计运用了抽象"山水"具象"庭园"结合的概念，设计精心构思出十栋高低错落、形态各异的三至七层的房子，围合成一建筑群。设计总体构思为：山外山，园中园。建筑空间设计体现岭南建筑空间自由、流畅、开敞的风格。建筑色调淡雅，造型富有新意。将传统的居住模式和现代生活结合。

建筑规划充分利用土地资源、保护现有的自然环境资源。设计依地势和空间的围合要求，盘旋而出一段山形般波折起伏的建筑形体，把基地环抱其中，将基地周围尤其是东面的厂房杂乱、烦杂摒弃在外，体现"山外青山楼外楼"的空间意向。建筑围合而成了一个大园子，园子中凿咫尺小池为镜，以桥为舟，一个个房子从"建筑山"生长出来，临水而居，营造出自然生态的岭南水乡景色。在大园中设计又勾画出与大园相互连通的六个小庭院，它们各具不同的特色，尺度宜人，景物少而精，有的庭中栽竹，形成竹园；有的植松，种梅，供朝夕观赏，有的庭院与大园中水面相临，用桥、廊、墙将水池一湾围合成小院，空间流通，层次丰富，景色生动，做到移步换景，希望能把桃花源的意境嫁接到现代城市的生活当中去。

Maillen Apartment is in Banshan area, Shekou, Shenzhen, with Yan Shan Lu on the east, Banshan Seaview Villas on the west, and Banshan community center and Lanxigu high-end residential quarter on the north. Shekou is situated in the southeast of Nantou Island in Shenzhen with Shenzhen Bay on the east and the Pearl River Estuary on the west. Overlooking the west part of Hong Kong across the sea, it is the forefront for China's reform and opening up. More than 3 decades of reform and opening up in economic development has transformed Shekou into an international community with nearly a hundred world-renowned enterprises and a most internationally compatible place in Shenzhen. Maillen Apartment is envisioned as a high-end serviced apartment offering small units for high-caliber talents from abroad.

With the increasingly generalized and assimilated overall urban presence as result of globalization and commercialization, the design of Maillen Apartment intends to create a residential mode that highlights localization and individualization. Abstract mountains and rivers are combined with the concrete courtyards in the meticulous design of ten high and low buildings of three to seven floors in varied forms that embrace each other into a building complex. The general concept is mountains beyond the mountains and gardens inside the gardens. The design of architectural spaces embodies the free, fluid and open Lingnan style. With elegant light tones and novel building forms, the project represents a perfect combination of traditional residence and modern life.

Land resources are fully utilized and existing natural and environmental resources conserved in building planning. Based on landform and space enclosure requirements, a mountain-like undulating building volume winds its way, enclosing the site and separating messy factories and disorder on the east to create the image of *mountains beyond the mountains and buildings outside the buildings*. The buildings circle around a generous garden where the boat-shaped bridges span over the mirror-like pools. Buildings grow from the "mountain of buildings" near water, creating a natural and ecological landscape typical of Lingnan water village. Inside the large garden are six small courtyards of varied features and appropriate scale, all connected with the large garden and designed with a few but very gorgeous sceneries, some planted with bamboos into a bamboo garden, some planted with pine and prune trees and some adjacent to water in the large garden and enclosed by bridges, corridors and walls, all with interactive spaces, multiple layers and vivid sceneries. Every step is an encounter with a different scenery, all in an effort to blend the idea of Peach Blossom Paradise into urban life.

项目地点：广东省深圳市南山区蛇口
设计时间：2006-2008
建设时间：2008-2010
建筑面积：25142m²
建筑层数：3-7层
建筑高度：12.35/23.45m
合作单位：深圳市都市实践设计有限公司
曾获奖项：2011年度广东省优秀工程勘察设计奖（住宅类）一等奖
2012年广东省岭南特色建筑设计奖铜奖

Location: Shekou, Nanshan District, Shenzhen, Guangdong Province
Design: 2006-2008
Construction: 2008-2010
GFA: 25,142m²
Floors: 3-7
Height: 12.35/23.45m
Partner: URBANUS
Awards: The First Prize of Excellent Engineering Exploration and Design Award (Residence) of Guangdong Province, 2011
Bronze Award of Lingnan Feature Architectural Design Award of Guangdong Province

1. 庭院与大园中水面相临，用桥、廊、墙将水池一湾围合成小院，空间流通，层次丰富，景色生动
   The courtyard is adjacent to water in the large garden, enclosed by bridges, corridors and walls and presenting interactive spaces, multiple layers and lively views

2. 建筑围合而成了一个大园子，园子中凿咫尺小池为镜，以桥为舟，临水而居，营造出自然生态的岭南水乡景色
   The buildings circle around a large garden where small ponds look like mirrors, bridges are like boats, and life is near water, creating a natural and ecological landscape typical of Lingnan water village

3

4

5

3 建筑依地势和空间的围合要求，盘旋而出一段山形般波折起伏的高低错落、形态各异的三至七层的房子，围合成一建筑群
Based on landform and space enclosure requirements, a mountain-like undulating building shape winds its way into high and low buildings of three to seven floors in varied forms that embrace each other into a building complex

4 建筑依山而建，把基地环抱其中，体现"山外青山楼外楼"的空间意向
Buildings are constructed near mountains, enclosing the site for the intention of creating "mountains beyond the mountains and buildings outside the buildings"

5 小庭院空间尺度宜人
Proper size and pleasant space of small courtyard

6 独特的建筑造型
Unique building style

7 建筑空间设计体现岭南建筑空间自由、流畅、开敞的风格
The design of architectural spaces embodies the free, fluid and open Lingnan style

8 建筑色调淡雅，造型富有新意
Light building colors and innovative building forms

9 相互连通的庭院空间
Interconnected courtyard spaces

广州科学城科技人员公寓
S & T Professionals' Apartment, Guangzhou Science City

广州萝岗新城是广深经济走廊上的科研孵化中心、广州东部地区的现代化服务中心。为吸引海外留学人员归国创业，满足在广州萝岗新城工作的外籍人士、留学归国人员及科技人员、专家学者等住宿的要求，广州萝岗区启动了广州科学城科技人员公寓的建设项目。

建筑采用围合式布局，两栋塔楼公寓和两栋板式多层公寓围合出内部庭院。庭院内布置有会所建筑与泳池、下沉花园等公共活动空间。科技人员公寓的定位、服务对象及标准多样化，住宅单元类型多。在保证居住空间私密性的前提下，设置促进邻里交往的公共空中花园。通透的空中花园、与公寓绿化阳台、自由组合的遮阳百叶构成了独特的、富有灵气的建筑形象。两栋塔楼采用微纺锤形的剖面设计，成为当地的标志建筑。平面周边的柱子上下端采用斜柱，中段直柱连接的形式，并采用直线预应力钢筋解决转折处楼面构件受拉问题，避免建筑结构的竖向不规则的不利影响。

科技人员公寓主要采用了被动式太阳能设计。居住空间坐北向南，间距适中，适应岭南地区的气候特点；架空层和空中花园能够形成顺畅的风之通路；绿化露台与竖向百叶相互作用，达到有效的隔热效果，保证舒适的室内环境。另外，项目还使用了大量能减少对环境带来负荷的新技术、新设备，以实现绿化环境的目标。如雨水回收系统、冷交换热泵供水系统、节水洁具、智能电气系统、智能多联体空调系统、LED节能园林灯等，有效地降低运营费用。

Luogang New Town serves as an S&T incubator in Guangzhou-Shenzhen economic corridor and a modern service center in the eastern part of Guangzhou. The Project was launched by the government of Luogang District in a bid to attract overseas students back to China for entrepreneurship and to address the accommodation needs of expatriates, returned overseas students, S&T professionals and experts working in Luogang New Town.

Based on an enclosing layout, two tower apartments and two slab-type multi-floor apartments form an inner courtyard, inside which spaces for public activities such as clubs, swimming pools, and sunken gardens are provided. The Project is planned with diversified positioning, service objects and standards as well as various apartment types. While ensuring the privacy of the residential space, public sky gardens are provided to enhance neighborhoods interaction. Transparent sky garden, greening terrace of apartment and freely combined sun shading shutters form a distinctive and vivid architectural image. The two towers with fusiform section are made locally representative buildings. The columns in the perimeter of the plane feature inclined columns at the upper and lower ends connected by straight column in the middle section. Straight prestressed tendon is employed to address the tension of building components at the turning point and avoid the adverse impact caused by the vertically irregular structure.

The Project adopts passive solar design. The apartment units facing the south and being moderately spaced perfectly adapt to the climate characteristics of Lingnan region; a smooth ventilation passage is formed between the open-up floor and the sky garden; the greening terrace and vertical shutters interacting with each other ensure effective insulation and comfortable indoor environment. In addition, multiple new technologies and equipment are applied to minimize environmental loads and foster greening environment. Rainwater recycling system, heat pump water supply system, water-saving sanitary ware, smart electrical system, smart multi-connected air conditioning system, and LED energy-saving garden lights are employed to effectively cut the operation cost.

项目地点：广州市萝岗区
设计时间：2006-2007
建设时间：2008-2010
用地面积：39957m²
建筑面积：105358m²
建筑层数：22层
建筑高度：77.75m（塔楼），31.3m（板式公寓）
合作单位：日本佐藤综合计画
曾获奖项：2006年国际竞赛第一名
　　　　　2015年度全国优秀工程勘察设计行业奖 一等奖
　　　　　2011年第六届中国建筑学会建筑创作优秀奖
　　　　　2011年广东省注册建筑师优秀建筑创作奖
　　　　　2011年度广东省优秀工程勘察设计奖二等奖

Location: Luogang District, Guangzhou
Design: 2006-2007
Construction: 2008-2010
Site: 39,957m²
GFA: 105,358m²
Floors: 22
Height: 77.75 (tower), 31.3m (bar-type apartment)
Partner: AXS
Awards: The First Place of international competition in 2006
The First Prize of National Excellent Engineering Exploration and Design Award, 2015
Excellent Award of the Sixth ASC Architectural Creation Award, 2011
Excellent Architecture Creation Award by Guangdong Chapter of Association of Chinese Registered Architects, 2011
The Second Prize of Excellent Engineering Design Award of Guangdong Province, 2011

1　总平面图
　　Site plan

2　纺锤形双塔成为萝岗区标志性建筑
　　The spindly twin towers are made the landmarks in Luogang District

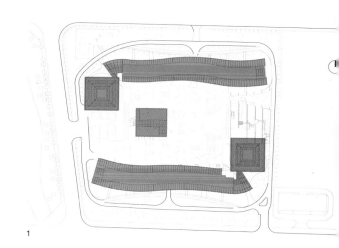

1

3　在保证居住空间私密性的前提下,设置促进邻里交往的公共空中花园。通透的空中花园、与公寓绿化阳台、自由组合的遮阳百叶构成了独特的、富有灵气的建筑形象
　　While ensuring the privacy of the residential space, public sky gardens are provided to encourage neighborhoods interation. Transparent sky garden, greening terrace of apartment and freely combined sun shading shutters form a distinctive and vivid architectural image

4　板式公寓:巧妙利用板式公寓空中花园两侧的山墙面面,采用鲜艳色彩的外墙涂料,赋予每一个住宅组团独特的个性
　　Slab-type apartment: The gable walls on both sides of the sky garden of the slab-type apartment are coated in bright color, giving unique individuality to each residential cluster

5　建筑立面细节:垂直百叶、屋面遮阳百叶及大进深露台相互作用,保证整个建筑的节能效果
　　Facade details: The interaction between vertical shutters, rooftop sun shading shutters and large-depth terrace ensures the energy efficiency of the entire building

6　富现代科技气息的内庭园景
　　Inner courtyard view with modern scientific and technological atmosphere

7　多层板式公寓利用顶层的钢架与高层塔楼铰接相连,以增强结构的稳定性。板式公寓末端采用V形钢管混凝土斜柱支撑从下至上逐渐增大的绿化平台
　　Multi-floor slab-type apartments are hinged with the high-rise tower via the steel structure on the top level to further stabilize the structure. At the end of the slab-type apartment, V-shape steel-pipe concrete tilted-columns are adopted to support the green platform gradually enlarged upward from bottom

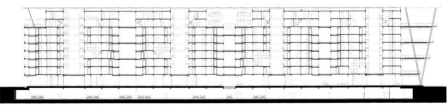

Residential Buildings 居住建筑　235

佛山依云水岸A1、B1、B2项目
Project A1, B1 and B2 of Evian Town, Foshan

佛山新城的功能定位为佛山市商贸金融、文化体育、休闲娱乐和信息服务中心、高品质生活环境的城市生活区。佛山依云水岸A1、B1、B2项目即位于佛山新城，项目的景观资源丰富，地块北面是东平河，在河岸边是大型沿岸绿化公园，地块西面为未建城市保留绿地，地块东南方向为中心区的中央公园。项目目标客户群为本地高端客户群，项目已成为佛山市顶级TH社区、高尚住区的代表。

规划设计中，为充分利用地块周边景观资源，尤其是北侧江景，争取最大的景观朝向，高层布置时采用的是开放式的手法，合理设计视线与各种流线，达到对外部环境资源的有效利用，以及自身与环境相互协调的目的。建筑组合群体具有相对的整体性，形成良好的群体形象及完整的外部空间形态的同时，作到户户有景观。

建筑设计中通过楼盘外部整体形象的烘托，力求为本地区树立全新的形象。立面风格是创新性的新古典风情，整体外观华贵。户型设计注重灵活性、趣味性，以体现景观住宅特色，例如部分户型可提供大面积的阳台、空中花园等手法，以适应和引导不断增长的个性化需求，每一种户型都应有其独特设计亮点。

Foshan New Town is planned as a center for business and finance, culture and sports, entertainment and information services in Foshan, and an urban community with quality living environment. Located d in Foshan New Town, the Project enjoys rich landscape resources, neighboring Dongping River to the north with a large greening park by the riverside, reserved green spaces for urban development on the west, and the central park of the central area to the southeast. The Project targets at high-end local customers and has already become an icon for top TH community and high quality residential community in Foshan.

During the planning and design, in order to make full use of the peripheral landscape resources, especially to obtain maximum view into the river on the north, an open layout was employed for the high-rise buildings. Sightline and various circulations were reasonably designed while external environment resources were efficiently utilized to coordinate the Project with the environment. The ensemble of buildings shapes up a good cluster image and integrated external form, while ensuring a lovely view for every household.

By highlighting the overall image of the buildings, the architectural design seeks to establish a brand-new image for the local communities. The facades are designed in innovative neo-classic style, presenting a luxurious overall appearance. The apartment types are designed in a flexible and interesting manner to reflect the features of a residential community with a view. For example, some units may be provided with generous balcony or sky garden to meet the growing individualized demands of target customers. Each apartment type is ensured with distinct design highlights.

项目地点：佛山
设计时间：2007-2008
建设时间：2008-2012
建筑面积：520000m²
建筑层数：42层
建筑高度：129.3m
曾获奖项：2007年度广东省优秀工程勘察设计三等奖

Location: Foshan
Design: 2007-2008
Construction: 2008-2012
GFA: 520,000m²
Floors: 42
Height: 129.3m
Awards: The Third Prize of Excellent Engineering Exploration and Design Award of Guangdong Province, 2007

1 总平面图
　Site plan

2 高层区沿河景观：沿河建筑呈曲线布置，与河道形成有机呼应，体现建筑群刚中带柔的美感
　Riverside Highrises: the curvilinear layout of the waterfront buildings echoes to the river form and brings gentleness out of the masculine buildings

3 别墅区鸟瞰效果：立面色彩采用招商地产经典的"红、白、灰"组合
　Bird's eye view of the villa area: façade in "red, white and gray", a classical combination used by China Merchants Property

4 双拼别墅：立面风格干净利落，体现整体大气形象
　Semidetached villa: Clean and neat façade presenting a generous image on the whole

5 公共下沉式庭院：公共空间强调的是邻里间休憩交流的自然发生
　Public sunken courtyard: Emphasizing the spontaneous and natural communication between neighbors in public space

4

5

香港新福港地产·广州萝岗鼎峰
SFK · DF Project, Luogang, Guangzhou

广州市从亚运会前夕启动了全市范围的三旧改造进程，旨在更有效、更合理的利用有限的城市土地，创造更好的城市生活环境。本项目性质也属于此类。项目选址位于萝岗区政府附近，北侧为善坑山，南侧为城市主干道，面向广州演艺中心及拟建的商业综合体，附近是地铁14号线萝岗站，地理位置十分优越。"因地制宜，筑半山好住宅；融会贯通，造岭南新社区"是本项目设计的重要宗旨。

项目包含回迁区及融资区两部分。回迁区位于线坑村原址上，是线坑村民的回迁住宅及出租物业。在布局上，充分考虑了当地的风俗生活，布置了满足祭祀庆典的祠堂广场，及各种小尺度的邻里交往空间，使村落的习俗文化得以延续。融资区位于善坑山山腰，是面向房产市场的中高档商品住宅。设计充分利用山地地形，创造高低错落，空间体验丰富的住区环境；建筑布局疏密有致，尊重城市轴线，延续城市视廊，使山景融入社区及城市景观中。

整体建筑造型简洁现代，体现时代特色。住宅顶层采用悬挑的百叶屋架，既起到屋面隔热的作用，又能体现岭南建筑轻盈的神韵。两区的建筑造型整体色调一致，细部上有所区别，平衡了投资造价、住区氛围营造等多个方面因素。

项目地点：广州 萝岗中心区
设计时间：2011
建设时间：2011年至今
用地面积：120000m²
建筑面积：约420000m²
建筑层数：1-28层
建筑高度：99.9m

Location: Luogang Central Area, Guangzhou
Design: 2011
Construction: 2011 to date
Site: 120,000m²
GFA: about 420,000m²
Floors: 1-28
Height: 99.9m

Right before the Asian Games 2010 Guangzhou, the city launched the Initiative to redevelop old towns, old factories and old villages to maximize the use of limited urban land and create a better urban living environment. This project falls within this category. The site is located near Luogang District Government, with Shankeng Hill on the north and urban main road on the south, facing Guangzhou International Sports & Performing Arts Center and the planned commercial complex. The site is close to Luogang Station of Metro Line 14, enjoying a favorable geographical location. The design philosophy is to develop quality hillside residences and create new Lingnan-style community by incorporating site conditions and referencing the traditional architecture.

The project comprises the resettlement housing area and the developer-financed residential development. The former is located at the original place of Xiankeng Village, involving villagers' resettlement housing and rental properties. In layout design, we fully consider local customs and life styles, incorporate temple square for worshipping and celebration, and provide various small spaces for communications to carry on the village's customs and cultures. The developer-financed residential development located on the hillside of Shankeng Hill is to develop the market-oriented medium- and high-end commercial residences. The design makes best use of the hilly terrain to create staggered buildings and a residential environment of varied spaces, as well as an elaborately devised architectural layout, showing respect for the city axis, extending the visual corridor and bringing the mountain view into the community and the city.

The building forms are, in general, simple and modern, reflecting the features of the times. The top floors of buildings are installed with cantilever louver roofs, which not only serves the purpose of insulation but also reflect the ethereal grace of Lingnan architecture. The buildings in both areas generally share the same color tone with certain differentiated details, well balancing various factors including investment, residential ambience etc.

1

2

1　总平面图
　　Site plan

2　钢琴会所滨水而设，营造高品质的半山园景
　　The waterfront piano club contributes to the high quality hillside landscape

3　园区入口设置在香雪大道，郁郁葱葱如森林般，彰显小区的现代与时尚，静谧与祥和
　　The entrance is set on Xiang Xue Da Dao, where the forest-like greenery makes a stylish, tranquil and peaceful modern community

4　宁静端庄的入口
　　Tranquil and elegant entrance

5　灵动时尚的会所中庭
　　Trendy and Inspiring club atrium

招商伍兹公寓
China Merchants Woods Park

在海上世界中心地带兴建的伍兹公寓，山海美景，尽览无余，占据天时地利，以超光感外墙、陶土板等主要材料，打造地标式豪宅，诠释深圳豪宅新定位。

本项目的布置将3栋住宅塔楼设计成两梯两户的点式，沿地块西北-东南向错位布置，每两栋住宅塔楼之间的间距均大于13米，各栋住宅四个朝向的景观均无遮挡且每户至少有一个客厅或卧室朝向东面的海景，每栋住宅的每一户都能获得优美的山景及海景景观，不影响从大南山到海岸线的绿色视觉走廊的通透性。主体采用了大板结构，使室内空间无梁少柱，增加了室内空间间隔的灵活性，保证了室内空间开敞、大气。除此之外，全部户型采用两层高超大露台的方式使每户均享有舒适的超大观景平台，提高了整个楼盘素质。

在外观设计上，建筑基座的表面纹理和上部的互相辉映衬托，基座坚实浑厚，采用较深的凹进，塔楼主体则营造清晰、流畅，水平突出向外伸出的阳台。外墙主要采用陶土板和玻璃幕墙，陶板具备传统土制材料的外观与透明玻璃之间的细微平衡赋予建筑无惧时间流逝的特质，营造恒久稳固之感。整个建筑选材和色彩以明快的浅色调为主，间以深色的实墙面，典雅大气，细部处理精致到位，彰显出高尚住宅的品质。

Built at the center of the Sea World, China Merchants Woods Park enjoys prominent location with unparalleled sea and mountain views. The cladding of super sheer exterior wall and ceramic panel help make the project a benchmark and new standard for luxury buildings in Shenzhen.

In layout, three residential towers in the park are designed into towers with two units and two elevators. The towers are staggered along the northwest-southeast direction, with the interval of over 13m between every two residential towers. Each residential tower enjoys unobstructed views on all four directions, and each unit has at least one sitting room or bedroom facing the sea on the east. Each unit in each residential tower enjoys beautiful mountain and sea view, without affecting the transparency of the green visual corridor from Danan Mt. to the seashore. The large panel structure help create beam-free space with minimal columns and enhance the flexible interior partitioning, providing an open and spacious internal space. Besides, all units come with a two-floor-high large terrace which offers generous and comfortable observation deck while upgrading the overall quality of the entire project.

In terms of building appearance, the surface texture of building base echoes with that of the upper part, complementing each other. The base is solid and thick with deep recesses, while the tower is distinct and fluid with horizontally protruding balconies. The facade mainly consists of ceramic panel and glass curtain wall. The ceramic panels present the visual quality of traditional earthen materials, and when balanced with transparent glass, contribute to a timeless and stable building image. The materials and colors for the building are primarily light tones with intermittent dark solid walls, expressing elegance and grace, and the fine details also interpret the remarkable quality of a high-end residential development.

项目地点：深圳市南山区蛇口海上世界
设计时间：2010
建设时间：2010-2012
建筑面积：47211.4m²
建筑层数：28层塔楼和3层裙房
建筑高度：88.60m
合作单位：美国SBA建筑设计有限公司
曾获奖项：2013年度广东省优秀工程勘察设计奖住宅类一等奖
2013年度广东省优秀工程勘察设计奖一等奖
2014年中国建筑学会建筑创作奖居住类银奖

Location:	Sea World, Shekou, Nanshan District, Shenzhen
Design:	2010
Construction:	2010-2012
GFA:	47,211.4m²
Floors:	28 (Tower), 3 (podium)
Height:	88.60m
Partner:	Steffian Bradley Architects
Awards:	The First Prize of Excellent Architectural Design (Residence) of Guangdong Province, 2013
	The First Prize of Excellent Engineering Exploration and Design Award of Guangdong Province, 2013
	Silver Award of Architectural Creation Award (Residence) by the Architectural Society of China, 2014

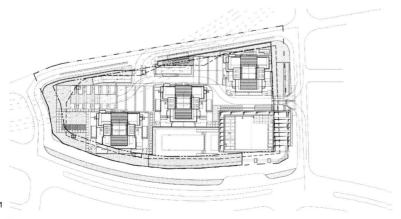

1

1. 屋顶总平面
   Roof plan

2. 超光感外墙，打造地标式豪宅
   Super sheer exterior wall contribute to creating a luxurious landmark.

3. 小区主入口
   Main entrance to the park

4. 背面陶土板百叶，虚与实相结合
   Ceramic plate louver on the back of the building is a combination of the void and the solidness

5. 错层阳台设计，独具韵律感
   Staggered balconies present a unique sense of rhythm

鲸山花园九期
Jingshan Garden Phase IX

蛇口大南山半山兴建的鲸山花园九期，毗邻蛇口国际滨海休闲片区。这里由于地势较高，于山林层岩处便能远眺大海；往南海的方向，又能俯瞰繁华都市。创作从"绿色建筑设计、自然生态美学、建筑类型学思考"三方面入手，继承传统村落依托山地有机成长的空间组织和风貌营建模式，用现代语汇设计宜居的高尚住区。通过巧妙的地形利用，鲸山九期花园的建筑布局依山、观海、瞰城，住区实现了"融入林泉、依山面海，居于半山之上、层楼掩映于林"之景。

设计建立在"绿色建筑"的视角上：尊重原有的地形地貌，保护原生的树木岩石景观。建筑布局依山就势、以西高东低的规划形态，其中最特别的是"爬坡住宅"，在竖向上发挥了上下承接的作用，实现了对地形的最大尊重。住区居住生活的各项内容（居住、交通、休闲）与自然生态系统融为一体，人们感受到亲近自然的乐趣，一种优雅的、介乎自然与人工的山居生活跃然而出。

在设计上，将建筑与景观巧妙融合，仿佛建筑是自然生长出来的，我们选择了立方几何形体和方框形构件为造型母题，立方几何形体呼应独特的岩石形态，错层式的立方阳台突出于大面墙体，米黄色陶板等自然的材料运用与金属铝板形成强烈对比，雕琢出质朴、内敛、简约的住宅建筑内在本质。外立面中的方框形构件元素透露出"家园"的隐喻，也成为住区场所氛围构成的核心词汇。方框形构件的"方"构形，源自于岭南传统民居三间两廊的"方院"，是深入建筑类型学思考的成果。方框形构件模块组合、叠合，表述一种"自然生长"的意义。

Jingshan Garden, Phase IX is situated halfway up Danan Mt. in Shekou, adjacent to Shekou seaside leisure area. With a relatively large difference in elevation, the garden enjoys sea view from mountain forests and rocks, and overlooks the bustling metropolis in the direction of the South China Sea. The design of the garden, from the perspective of three aspects, i.e., "green building design, natural eco-aesthetics, and building typology", inherits space organization and landscape development mode of traditional villages which focus on organic growth with the mountains, creating a livable high-end residential building design through modern architectural vocabulary. Through ingenious utilization of terrain, the garden nests in the mountain, facing the sea and overlooking the city, presenting a picturesque view of the residential district where the *buildings at halfway up the mountains are nestle amid the trees, springs, mountains and sea.*

The Garden is designed from the perspective of green building: respect existing landscape, and protect natural treescape and rock formation. In architectural layout, the buildings, nesting in mountains, are oriented in a way to comply with the planned building layout, i.e. higher in the west and lower in the east. The most special are the hillside buildings, which function as the connection between the upper and lower parts, showing full respect to existing landscape. Life in the residential district (residence, transportation and leisure) is naturally integrated with the eco-system, where people can enjoy the fun of being close to nature, highlighting vividly an elegant mountain lifestyle right between what is natural and artificial.

In the design, buildings are delicately blended into landscape as if naturally growing into existing form. Solid geometric shapes and box-shaped components are employed as motif pattern. The solid geometric shapes echoing with unique rock formation, the staggered cubic balconies protruding from the large wall, and the natural materials including beige ceramic plate etc. in sharp contrast with the metal aluminum plate are meticulously constructed to exhibit the intrinsic nature of a plain, modest and minimal residential building. The box-shaped components on facade are a metaphor for home, and also a key word for the ambience in the residential district. The square shape of the component is originated from the quadrangle with three houses and two corridors in traditional Lingnan houses, achieved through in-depth study of architectural typology. The box-shaped component modules are assembled and stacked to present a setting of natural growth.

项目地点：深圳市南山区蛇口鲸山观海
设计时间：2010
建设时间：2010-2014
建筑面积：165145.89m²
建筑层数：28层
建筑高度：90m
合作单位：美国SBA建筑设计有限公司
曾获奖项：2015年度全国优秀工程勘察设计行业奖一等奖
　　　　　2015年度广东省优秀工程勘察设计奖一等奖
　　　　　深圳市第十六届优秀工程勘察设计（住宅建筑）二等奖

Location:　　　Jingshan Jinhai, Shekou, Nanshan District, Shenzhen
Design:　　　　2010
Construction:　 2010-2014
GFA:　　　　　 165,145.89m²
Floors:　　　　 28
Height:　　　　 90m
Partner:　　　　Steffian Bradley Architects
Awards:　　　　The First Prize of National Excellent Engineering Exploration and Design Award, 2015
　　　　　　　 The First Prize of Excellent Engineering Exploration and Design Award of Guangdong Province, 2015
　　　　　　　 The Second Prize of the Sixteenth Excellent Engineering Exploration and Design Award (Residence) of Shenzhen

1　总平面图
　　Site plan

2　鸟瞰图
　　Bird's eye view

3　高层住宅
　　Residential high-rise

4　配套商业会所
　　Supporting business club

5　联排流水爬坡——整体框架置身于坡地递退的形态中，强调用舒展的水平线条来勾勒理性的地形面貌，利用悬挑结构使大体量分解成轻盈的块体，使建筑带有超脱重力的雕塑感，外墙材料以抽象植物肌理的印刷混凝土和印刷玻璃为主，以质朴平实的姿态与坡地对话
　　Sloping townhouses - the overall frame is incorporated with the setback of the terraces. The concept is to outline the rational terrain with the flowing horizontal lines, and divide massive building into light building parts with the overhanging structure. Thus a gravity-free and sculpture-like building appearance is created. The facade features the printed concrete and printed glass with the texture of abstract plant patterns, establishing dialogue with the site in a humble and modest manner

6　剖面图
　　Section

2
3
4

5

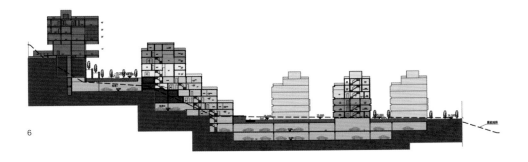

6

Residential Buildings 居住建筑

7

7 广宙叠梯住宅 ——"广宙建筑",即一种完全分散的、低密度的生活形式,一种与自然环境更加贴近的田园生活氛围。开放式的屋顶绿化平台有利于增加空间层次感,与自然的坡地和园林景观相得益彰
Stacked housing represents a totally distributed and low-density life style and a idyllic environment that is closely tied to the natural environment. The open green roof enhances the sense of spatial hierachy and harmonizes with the natural sloping terrain and landscape

8 独栋悬挑住宅剖面
Section of free-standing overhung house

9 高层入口一景
View of high-rise entrance

10 独栋悬挑住宅
Free-standing overhung house

11 联排流水爬坡
Flowing townhouse

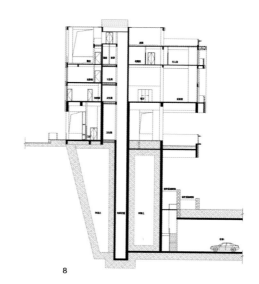

8

9

10

11

Residential Buildings 居住建筑   245

万科科学城项目—— 一期工程
Vanke Project in Guangzhou Science City (Phase I)

广州萝岗新城是广深经济走廊上的科研孵化中心、广州东部地区的现代化服务中心。项目位于广州萝岗新城，开泰大道与瑞和路交汇处的东南侧。规划用地177588平方米，总建筑面积为57万平方米，计容建筑面积为44万平方米，规划容积率为2.5，规划人口约三千人。

项目包括高层住宅、小学、幼儿园、商业、卫生所等，是集居住、教育、商业于一体的大型生态居住社区。项目通过以下几点，诠释了新时代科技背景下新中式岭南风格的丰富内涵：1、住宅单元共性的标准化设计与配套设施个性化追求；2、岭南风格的现代传承；3、工业化技术的探索性应用。

万科东荟城项目通过对中国传统文化与美学的继承与发展，营造出健康、充满活力、富于文化的现代"新中式"优质社区。

Luogang New Town serves as an S&T incubator in Guangzhou-Shenzhen economic corridor and a modern service center in the eastern part of Guangzhou. Located southeast of the intersection of Kai Tai Da Dao and Rui He Lu in Guangzhou Science City, Luogang District, Guangzhou, Vanke Project (City Gate) is planned with a GFA of 570,000m^2 and a FAR-counted GFA of 440,000m^2 on a site of 177,588m^2, a FAR of 2.5 and a planned population of 3,000 people.

As a large ecological residential community integrating residence, education and retail, the project includes residential high-rises, elementary school, kindergarten, retail and health center etc.. The project interprets the rich connotations of the neo-Chinese Lingnan Style at the age of new technology in the following aspects: 1. standardized design of residential units and individualized supporting facilities; 2. inheritance of Lingnan Style in modern society; 3. exploratory application of industrial technology.

The project, through the inheritance and development of traditional Chinese culture and esthetics, creates a healthy and dynamic neo-Chinese modern community of cultural connotation.

2

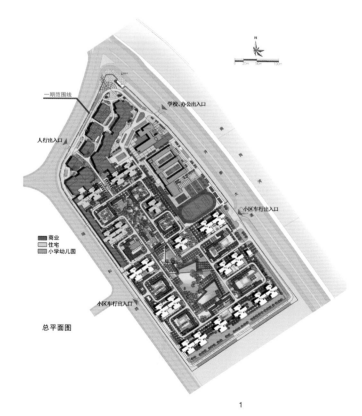

1

项目地点：广州市萝岗区科学城开泰大道与瑞和路交界处
设计时间：2009-2010
建设时间：2010-2015
建筑面积：177588m^2
建筑层数：33层
建筑高度：95.8m
合作单位：思邦建筑设计事务所
曾获奖项：2015年度广东省优秀工程勘察设计奖一等奖

Location:	Crossing of Kai Tai Da Dao and Rui He Lu, Science City, Luogang District, Guangzhou
Design:	2009-2010
Construction:	2010-2015
GFA:	177,588m^2
Floors:	33
Height:	95.8m
Partner:	SPARK
Awards:	The First Prize of Excellent Engineering Exploration and Design Award of Guangdong Province, 2015

1 总平面图
Site plan

2 打造自由灵活的商业形态，犹如都市中心里的"自由村落"，营造放松、自由、舒适的办公与购物环境
Free-flowing and flexible business facility, just like a village of freedom within the urban context, offers a relaxing, free and comfortable office and shopping environment

3 万科东荟城项目位于广州萝岗科学城，项目包括高层住宅、小学、幼儿园、商业、卫生所等，是集居住、教育、商业于一体的大型生态居住社区
Located in Guangzhou Science City in Luogang District, Guangzhou, Vanke Project is a large ecological residential community integrating residence, education and retail, including residential high-rises, elementary school, kindergarten, retail and health center etc.

Residential Buildings 居住建筑 **247**

4 商业街景
View of commercial street

5 每户均有较好的朝向和风景,建筑间距在满足日照间距的前提下,更大程度的提高建筑物之间的距离,使组团之间有更大面积的庭院空间
Each unit enjoys favorable orientation and view. The distances between buildings, after meeting the sunlight requirements, are maximized for more generous courtyard spaces

6 园林小筑:传承岭南空间、岭南印象回忆
The gardening amenities represent the inheritance of Lingnan-styled space and the fond memories about the traditional Lingnan elements

7 住宅入口大门
Main entrance to the residential building

金隅大成 · 海口美灵湖住宅小区
Jinyu Dacheng – Meiling Lake Residential District, Haikou

本项目位于海南省海口市灵山西片区，在美灵湖边美兰机场附近。建设用地面积141940平方米；总建筑面积73679.07平方米（其中：住宅建筑70068.13平方米，会所商业楼3610.94平方米）。

海口地处热带，根据海口地区居民的住宅立面及平面布局特点，同时考虑当地居民的起居习惯，设计中注意"地理环境、园林、建筑"相互协调统一。利用小区内道路与室外水系的高差设置开敞的地下空间，将室外景观引入地下室，避免了一般地下室的单调黑暗。平面布局首层功能通过入户玄关引导，分隔客卧及厨房、客厅及厨房等空间，二层为客卧和主卧室，卧室均设有卫生间。每个房间均有良好的日照和通风。建筑效果设计在平面布局的基础上，运用海南热带建筑的特点布置坡屋面，利用檐口、阳台、走廊、露台、方木构件、灰土瓦及火山岩等多种元素之间的相互搭接，使建筑与周围环境相融合。

设计中将美灵湖与小区内水景相呼应设置无边界水池，使小区内外景观连为一体。如此设计景观，无限增添室外环境范围，提升小区档次。为了防止雨季美灵湖水位抬高和干旱对小区的影响设置了最高和最低警戒水位，水多时排涝。因本小区位于美兰机场飞机的飞行航线上，空中飞机噪声影响住户。根据机场提供的在此航线上飞机的噪声的数据，设置中空玻璃。小区内道路高于水面，设计中利用小区道路与水景的高差设置开敞的地下室，形成开放的地下空间。每栋住宅均根据不同的朝向设置屋面太阳能设施。

Located close to Meilan Airport beside Meiling Lake in western Lingshan area, Haikou, Hainan Province, this residential district is planned with a site area of 141,940m² and a GFA of 73,679.07m² (including: 70,068.13m² for residential buildings and 3,610.94m² for commercial and club facilities).

Haikou is in tropical region, so the typical features of façade and plan of the local houses and the living habits of local residents are considered in design to create harmony and unity of geographical environment, garden and architecture. With height difference between internal roads and exterior water system, an open underground space is created, bringing exterior landscape into the open below-grade spaces which would otherwise be dark and dull. On the first floor is a foyer leading the way into different functional spaces including separated guest bedroom and kitchen, sitting room and kitchen etc. On the second floor are guest bedroom and master bedroom, each with its own bathroom. Each room enjoys favorable daylighting and ventilation. The design of architectural effect incorporates, apart from organizing planar layout, sloped roof and various overlapped elements typical of tropical architecture in Hainan including the eave, balcony, corridor, terrace and square wood element, grey tiles and lava etc., making the architecture in harmony with the environment.

Water features in the district are designed into infinite pool to echo with Meiling Lake, combining the internal and external landscape into a unity. Such landscape design extending into the exterior upgrades the quality of this district. To minimize the impact of drought and water level rise in Meiling Lake, the highest and lowest warning levels are set to drain water in case of flood. As the district is located right below the flight routes of Meilan Airport and is affected by the aircraft noise, hollow glass is used according to the aircraft noise level provided by the airport. The internal roads in the district are higher than water level, and the height difference is utilized in design to create an open basement and underground space. Each residential building is equipped with solar energy equipment on roof in view of its orientation.

项目地点：海南省海口市灵山西片区
设计时间：2011
建设时间：2011-2013
建筑面积：73679.07m²
建筑层数：2层
建筑高度：7.8m
曾获奖项：2015年度广东省优秀工程勘察设计奖二等奖

Location: West Zone of Lingshan, Haikou City, Hainan Province
Design: 2011
Construction: 2011-2013
GFA: 73,679.07m²
Floors: 2
Height: 7.8m
Awards: The Second Prize of Excellent Engineering Exploration and Design Award of Guangdong Province, 2015

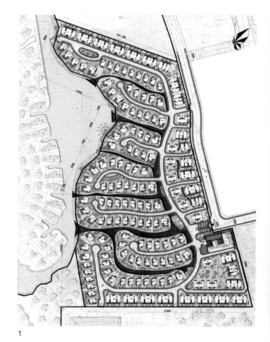

1　总平面图
　　Site plan

2　联排四拼B型住宅
　　Perspective of quadruple townhouse Type B

3　联排三拼B型住宅
　　Perspective of triple townhouse Type B

4　独立住宅A型
　　Perspective of free-standing house Type A

5　独立住宅B型
　　Perspective of free-standing house Type B

金山谷花园3B期、八期
The Hills Phase III-B and Phase VIII

金山谷花园2009年荣获联合国人居企业最佳范例奖,是中国唯一获此殊荣的项目。项目规划遵循招商地产"绿色地产开发模式"。项目位于广州市番禺区,15分钟直达珠江新城CBD。

金山谷花园3B期工程共由34栋住宅及2个独立变电所组成,住宅为双联、三联等多种形式。结合环境特点,营造山地建筑特色,以简洁平和的手法,组合穿插,利用天然石材、面砖、木材及钢材与玻璃,充分表达材料本身特性,形成轻盈舒展的造型特征。色彩以材料本色为基调;白、灰、木色、金属灰色为主调;以天然、质朴、现代的造型特征,来诠释坡地高档住宅的身份特色。金山谷花园八期为一类高层商住楼。地下两层,首层、二层为裙楼商业及配套公建。塔楼主立面体量由"L"、"T"以及"C"几何形体组合构成,塔楼之间有5层连体相接,造型上通过大跨度连体塔楼的形式寓意"金山谷之门"。两个塔楼形体对称又不完全对称,稳中有变。

金山谷花园八期项目属于结构类型特别复杂的钢筋混凝土高层建筑,需要结构超限审查。裙楼范围采用框架结构。在裙楼顶部塔楼范围设梁式转换,支托塔楼上部住宅的剪力墙。塔楼采用框支剪力墙结构。连体净跨度21.8米,柱中心跨度23米。连体部分采用轻骨料混凝土+组合结构组成的空腹桁架体系。本工程为了施工安全,施工阶段在连体的范围内采用贝雷架辅助支撑。从下而上由地下室底板直至连体16层检修层梁底。裙楼及地下室已建成的所有各楼层间用钢柱支撑,整个连体的施工荷载均由贝雷架传至底板。17-22层连体部分的所有荷载在连体结构强度形成前均由贝雷架支撑。

The Hills was awarded the Dubai Award for Best Practices to Improve the Living Environment 2009, being only winner of this award in China. Located in Panyu District, Guangzhou with just 15 minutes' drive to the Zhujiang New Town CBD, the Project is planned in compliance with the green property development mode of the China Merchants Property.

The Hills Phase III-B Project consists of 34 residential buildings and 2 individual substations. The residential buildings of multiple types including duplex, triplex etc., are interspersed and incorporated into the natural environment in a simple and tranquil manner, creating a characterized landscape of hillside buildings. The original properties of construction materials, including natural stones, tiles, wood, steel and glass, are fully retained to make the architectural style light and stretching. The dominant colors are the natural colors of the materials, primarily white, grey, wood color and metallic grey. The natural, plain and modern architectural style well demonstrates the identity of high-grade hillside residential buildings. The Hills Phase VIII is a project of Class One high-rise commercial and residential buildings. The two floors underground and the first and second floors aboveground comprise the podium building for retails and supporting public facilities. The tower buildings, with geometric shapes of "L", "T" and "C" making up the main facade, are connected by a five-floor long-span connecting structure to resemble the shape of "Gate of The Hills". The two tower buildings look symmetric but are not exactly the same, hence realize the diversity out of the unity.

The Hills Phase VIII Project involves high-rise buildings of very complex reinforced concrete structures exceeding code limits, and therefore requires expert review. The podium building is a frame structure. On top of the podium building within the tower building is a beam conversion to support the shear wall of upper residential tower building. The tower building is a framed shear wall structure. The connecting structure, a Vierendeel truss system of light aggregate concrete and composite structure, is 21.8m in net span and 23m between column centers. For the safety during construction, the connecting structure was supported by Bailey frame from the bottom slab of basement up to the beam bottom of the sixteenth maintenance floor. All the existing floors of podium building and basement will be supported by steel column, and the entire construction load of the connecting structure are transferred to bottom slab through Bailey frame. Before reaching required strength, the entire load of the connecting structure between the seventeenth and the twenty-second floor are supported by Bailey frame.

项目地点:广州
设计时间:2010.2-2011.7
建设时间:2012-2013
建筑高度:3B:建筑≤15m、八期:84.3m
建筑层数:3B:2-3层别墅、八期:21层、25层
曾获奖项:2015年度广东省优秀工程勘察设计奖一等奖

Location: Guangzhou
Design: 2010.2-2011.7
Construction: 2012-2013
Height: Phase III-B: Building ≤ 15m, Phase VIII: 84.3m
Floors: Phase III-B: 2-3 (villa); Phase VIII: 21/25
Award: The First Prize of Excellent Engineering Exploration and Design Award of Guangdong Province, 2015

1 总平面图
  Site plan

2 往山顶绿化公园方向整体外景
  Overall view towards the hilltop green park

3 别墅侧院
  Side yard of the villas

4 "金山谷之门"连体正立面
  Front facade of the connecting structure "Gate of The Hills"

5 从市广路南侧立面
  South facade viewed from Shi Guang Lu

3

4

5

华润小径湾花园（一期住宅及商业）
China Resources Xiaojing Bay Garden (Phase I Residence and Retails)

华润小径湾花园项目位于广东惠州市大亚湾区小径湾地区，项目南临小径湾海域，包含天然海湾，海岸线长约2公里，有大片天然优质沙滩；东、西、北三面群山环绕，中间有自然淡水小溪穿过，依山傍海，风景优美，包含住宅、五星级酒店、华润大学、商业街等多个项目。

一期住宅及商业项目位于小径湾花园东侧，南临海滩，设置住宅及商业功能，住宅小区建筑面积约31.7万平方米，地上32层，采用"点"、"线"、"面"结合的布局方式，形成一带、三个中心组团的小区空间。一线沿海布置退台式多层洋房，建筑尺度人性化，滨海空间舒适，点式高层住宅及板式高层弧线形布局，错落布置，形成良好前后进退关系，营造具有多维层次且具有变化的小区空间及沿海建筑界面，疏密有致，整个住宅区所有住宅都能享受到最大的海景。商业为多层建筑，建筑面积约9000平方米，地上3层，设置会所，商业及餐厅，造型整洁大方，沿海平面展开，层层退台，所有设备平面，立面出口及屋面高度作了严格把控。

项目设计采用现代滨海风格，简约整体，整体色调清新淡雅，采用水平线条与大海、沙滩相呼应。立面注重功能性，无装饰性构架，以确保住宅观海视线，颜色以滨海风格的白色为主，注重细部设计。

Located in Xiaojing Bay area of Daya Bay District in Huizhou, Guangdong, China Resources Xiaojing Bay Garden, facing the sea of Xiaojing Bay on the south, features natural bay, 2km coastline and vast natural and quality beach. Surrounded by mountains on the east, west and north and run through by freshwater creeks, the Project integrates the functions of residence, five-star hotel, China Resources University and commercial street with attractive view of mountains and sea.

In the east of the Garden and with beach on the south, Phase I is developed for the residential and commercial function. With the floor area of 317,000m² and 32 floors aboveground, the residential community is formed with one strip and three central clusters through the layout featuring the combination of "point", "line" and "plane". The setback multi-floor apartment buildings are provided along the 1st tier seafront area with human-scaled building size and comfortable coastal spaces. The point-type high-rise residential buildings and slab-type high-rise in an arch layout are well distributed to establish favorable spatial relationship and create multi-level and varied community spaces and coastal building interfaces, maximizing the sea view for all residential units. The multi-floor commercial buildings is designed with a floor area of about 9,000m² and 3 floors aboveground for club, retails and restaurants. The neat and elegant building is unfolded over the sea with cascading setback and strictly control over the accesses on MEP plan and elevation and roof height.

The modern coastal style looks concise and consistent with fresh and elegant tone. The horizontal lines echo with the sea and the beach. Dominated by the coastal white, the facade without decorative framework focuses on functionality and details to maximize sea view from residential units.

项目地点：惠州市霞涌小径湾
设计时间：2012.8
建设时间：2013.1
建筑面积：321100m²
建筑层数：地下1层，地上32层
建筑高度：100m
合作单位：澳大利亚柏涛设计咨询有限公司
曾获奖项：2017年度广东省优秀工程勘察设计奖一等奖

Location: Xiaojing Bay, Xiachong, Huizhou City
Design: 2012.8
Construction: 2013.1
GFA: 321,100m²
Floors: 1 underground, 32 aboveground
Building height: 100m
Partner: Australia PT Design Consultants Limited
Award: The First Prize of Excellent Engineering Exploration and Design Award of Guangdong Province, 2017

1 总平面图
　Site plan

2 整体日景
　Day view

3 庭院内景
　Internal courtyard view

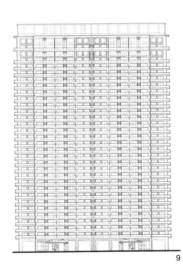

4　点式高层
　　Point-type high-rise

5　板式高层
　　Slab-type high-rise

6　庭院内景
　　Internal courtyard view

7　整体日景
　　Overall day view

8　点式住宅立面
　　Point-type residential building facade

9　板式住宅立面
　　Slab-type residential building facade

Residential Buildings 居住建筑　**257**

南沙港航华庭
Ganghang Huating Complex, Nansha

南沙港航华庭项目位于广州市南沙区，进港大道。用地周边有山体、高尔夫球场，北边有30米宽城市绿化带，东边距珠江约800米。环境条件十分优越。

广州南沙区是国家级新区和自贸试验区，进港大道是其主要的城市干道，属于城市景观重点地段。在项目设计及规划中，业主及区规划部门对建筑形象、建筑功能、空间关系等十分重视，希望本项目建成后能体现南沙新区的城市形象，并能成为重要的建筑景观。设计经历了多轮的方案研究比对，对空间布局、建筑功能规划、建筑立面风格、周边环境关系等都进行了多次优化调整，最后经由规划领导小组会议审议通过。本项目最终采用现代风格，高层住宅立面利用材质反差提高建筑物的辨识度；高层办公综合楼的最终使用单位为央企中铁港航局，配合其企业形象，建筑形体强调挺拔、稳重，立面采用现代主义设计风格，外立面的石材、金属构件与玻璃幕墙相互衬托，营造端庄大气，典雅精致的建筑形象。建筑形体主要采用高层点式布局，有效降低小区建筑密度，增加小区内园林和活动场地，提高城市空间的通透性，并与周边建筑高度相协调。设计结合岭南建筑的特点，如：强调住宅户型、内部空间的通透性，设置入户花园、主阳台、工作阳台，利用裙楼屋顶提供公共活动空间等。

本项目设置有屋面绿化、雨水回收利用系统，设计达到绿色建筑一星级的国家标准。

Located on Jin Gang Da Dao, Nansha District, Guangzhou, the Project is blessed with favorable environment, with mountains and golf course nearby, 30m-wide urban greening belt to the north and the Pearl River to the east within a distance of about 800m.

As Nansha District is defined as national new area and pilot free trade zone, while Jin Gang Da Dao, the main urban road, is an important part of city's landscaping system, the client and planning authorities paid great attention to the building presence, function and spatial relation and intended to make it a landmark to enhance Nansha's urban image. Following multiple rounds of analysis and comparison, the design scheme that has been optimized back and forth in terms of spatial layout, function programming, façade style and surrounding environment is deliberated and accepted at the leadership group meeting. The project eventually features a modern style. The high-rise residential building is highly recognizable thanks to the contrast of the facade materiality. In response to the corporate image of China Railway Port and Channel Engineering Group Co., Ltd, a central enterprise and the occupant of high-rise office complex, this towering and magnificent building is designed with modernist façade composed by stone, metallic components and glass curtain wall to exhibit a dignified yet elegant presence. The approach of point-type high-rises help effectively reduce the building coverage, spare more garden and activity spaces inside the community and enhance the openness of urban space. Moreover, it make the project harmonize with the surrounding buildings in height. The features of Lingnan architecture are also incorporated, such as putting more emphasis on the ventilation of apartment units and interiors spaces, providing in-house garden, main balcony and work balcony, and using the podium roof for public activity.

With roof greening and rainwater harvesting and reuse system, the project has met China's one-star Green Building standard.

项目地点：广州市南沙区进港大道
设计时间：2012.5
建设时间：2013.5
建筑面积：36088m²
建筑层数：地下1层，地上30层
建筑高度：99m
曾获奖项：2016年度广东省优秀工程勘察设计奖一等奖

Location: Jin Gang Da Dao, Nansha District, Guangzhou
Design: 2012.5
Construction: 2013.5
GFA: 36,088m²
Floors: 1 underground, 30 aboveground
Height: 99m
Award: The First Prize of Excellent Engineering Design Award of Guangdong Province, 2016

1 总平面图
  Site plan

2 办公楼主入口空间尺度合理，造型简练大方，符合央企形象
  Properly scaled and generously featured entrance of office building corresponds to the image of a central enterpirse

3 立面采用现代主义设计风格，端庄大气，典雅精致
  The modernist façade looks dignified yet elegant

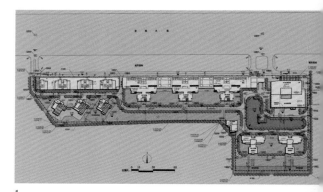

4/5 高层住宅形体方正，整体感强，立面利用材质反差提高建筑物的辨识度
The regular and holistic residential high-rise realizes high recognizability through the contrast of façade materials

6 住宅楼立面图
Elevation of residential building

7 高层住宅的远景
A distant view of residential high-rise

8/9 低层建筑轻巧舒展，与环境协调相融
The lively and extending low-rise well fit into the enviroment

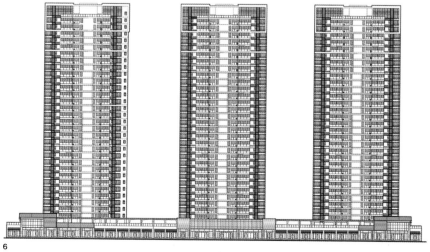

7

8

9

规划设计

作为大型综合设计院背景下的规划专业，GDAD规划团队依托院全面雄厚的多专业技术支撑优势，始终将"前沿视野、面向实施"作为我们专业的发展理念和核心特色。近年来，规划专业各团队在创作"可实施的规划"方面进行了一系列的积极思考和探索实践。

GDAD规划专业综合业务涵盖规划各种类型，其中特别需要多专业综合和讲求落地实施的详细规划，专项规划，村庄规划，规划咨询等类型规划为我院专长。目前，城市设计、TOD及地下空间规划、城市更新规划已成为GDAD规划专业的三大品牌业务。

注重国际视野与本地实际的结合，为项目提供国际视野，学习相关领域前沿经验，团队积极与DP，AECOM等国际知名顶尖团队在中新广州知识城、南沙咨询等开展良好合作。国外单位国际视野广阔，但先进理念实施落地是难题。本地团队对当地实际问题的理解与认知更为深入，与高水平国际团队合作，需要团队具有理念的共识与理解，当地问题的充分认知，优秀的表达转换能力及沟通协调能力。通过合理分工，各自发挥所长，以获得良好的项目效果。

实现多专业的高度融合，立足面向实施的发展理念，充分发挥GDAD的专业优势，以规划专业为统筹，组织道路交通、景观环境、公共服务设施、市政设施、建筑工程等各专业进行紧密合作，建立工作营模式，采用多元化的思维和多学科交叉的工作方法高效推进项目建设，如萝岗科学城和万博地下空间规划项目充分展示了GDAD多专业高度融合的强大优势。

探索"自下而上"的规划方法，通过更深的工作深度对详细规划进行推导验证，关注最终使用者需求及减少后续实施矛盾。如万博地下空间的工程反推、中新广州知识城及TOD等重要地块设计条件的建筑概念方案推敲论证等，以及猎德村改造项目中村民诉求在规划和建筑设计的落实，无不体现自下而上的技术优势。

坚持"规划-设计-建设-管理"一体化的规划理念，建立一体化的全面技术咨询团队，联动专业技术与管理运营，对规划建设项目实行全流程把控，建立长期动态的服务跟踪机制，为地区发展及规划项目高品质实施提供保障，并为规划管理部门科学决策及精细化管理提供重要支撑。南沙规划咨询项目作为广州首个整个城市范围的规划咨询探索，为团队的规划实践积累了重要的经验。

追求卓越的创新精神是团队不断进步的动力。团队在规划实践中紧贴国际视野、最新政策动向、行业前治、规划机遇，总结规划实践经验并结合项目实践创新规划方式，使团队不断积累实力与趋于向上。万博创新缓冲区概念、科学城探索以城市设计作为多规合一平台等都是创新探索的成果。

Planning and Design Urban

As an integral part of a large multi-disciplinary design institute, our planning team has been adhering to the development philosophy and core features of forefront vision and implementation orientation with great multidisciplinary supports from GDAD. In recent years, we conducted a series of proactive thinking, exploration and practice in terms of creating "enforceable planning".

Our planning services cover various planning typologies, having special advantage in detailed planning, specialized planning, village planning and planning consulting that particularly require multi-disciplinary coordination and implementation. At present, the urban design, TOD and underground space planning, and urban renewal planning have become the top three brand names of our planning service.

Combining the international vision with local situation has always been our focus. To undertake projects from an international perspective and learn the cutting-edge experience in related areas, our team has actively worked with internationally renowned counterparts such as DP and AECOM on the projects of Sino-Singapore Guangzhou Knowledge City and planning consulting project in Nansha. The foreign teams have broad international vision, but face the difficulty of implementing their advanced concepts; while the local teams have more in-depth understanding and perception of the actual local problems. Working with high-level international teams, the local team must develop conceptual consensus and understanding, full understanding of local problems, and excellent ability to express, communicate and coordinate. Through a reasonable division of works, all teams can play to their own strengths and achieve satisfactory project results.

Based on high-degree integration of multiple specialties, implementation-oriented development concept, and the disciplinary advantages of GDAD, we have been propelling close inter-disciplinary cooperation of road traffic, landscape, public service facilities, municipal facilities, and construction engineering within the planning framework. Workshops that incorporate diversified ways of thinking and interdisciplinary approaches are established to effectively advance the project, as evidenced by the planning of Luogang Science City and Underground Space of Wanbo CBD that fully demonstrate the strong advantage of our highly integrated specialties.

We have been exploring the bottom-up planning approach that deduces and validates detailed planning at a greater detailing level, and focuses on the needs of end-users to reduce conflicts in the subsequent implementation. The technical advantages of such approach have been proven by the backward induction in the planning project of the Underground Space of Wanbo CBD, the deliberation and validation of design conditions for the architectural conceptual design of important plots in Sino-Singapore Guangzhou Knowledge City and TOD development, as well as the incorporation of the villagers' requirements into the planning and architectural design of Liede Village Reconstruction Project.

Adhering to the concept of integrating planning, design, development and management, we aim to establish an integrated technical advisory team that coordinates expertise and management for a whole-process control over the operation of the planning construction projects. We have established a long-term dynamic follow-up mechanism to secure high-quality implementation of regional development and planning projects and provide essential support to planning authorities for scientific decision-making and fine management. The team has gained considerable experience from the Nansha planning consulting project, the first city-wide planning consulting attempt in Guangzhou.

The pursuit of excellence is the driving force for our continuous improvement. In our planning practices, we have been closely following the international visions, the latest policy trends, industry forefront and planning opportunities, summing up the experience in practice and innovating the planning approach based on the practice. In this way, we have been building up strength and making progress. The concept of innovation buffer area in Wanbo CBD and the idea of taking urban design as a platform for integrating multiple plans in the Science City are the results of our innovation and exploration.

中新广州知识城主城区城市设计深化及控制性详细规划
Urban Design Detailing and Regulatory Detailed Plan of Main Urban Area of Sino-Singapore Guangzhou Knowledge City

中新广州知识城位于广州市东部,是中国与新加坡两国共同倡导、推动的重大合作项目,是中新合作的第三代园区,将打造成为引领中国知识经济发展的新引擎。

知识城主城区位于知识城中部,面积32平方公里,将发展成为中新合作的国家创新型区域核心区、珠三角服务业对外开放创新区和科技金融中心、海内外优秀人才创新创业的重要基地、世界一流水平的生态宜居新城。本次城市设计以目标为导向,通过空间、经济、社会、文化、生态的整合,赋予知识城在区域中的特殊地点价值,营建珠三角地区汇聚智慧、统领发展、先导文明的空间场所,总体形成山水相融、簇团紧凑的城市空间特征。

本项目在规划创新方面:1)探索了城市设计落实为控规的方法和途径,将本次城市设计提升为规划管理的重要环节,有效提升城市空间品质;2)深度实践了"多规合一"的技术方法,确保规划"落地";3)构建起以图则、导则为载体的绿色生态指标落实途径,使"绿色生态"从理念走向实施。在规划实施方面:1)有效指导了知识城的市政基础设施、公共服务配套设施、景观工程等150余个项目的建设实施;2)有力地推动了知识城绿色生态城区的建设,助力知识城获批住建部国家智慧城市首批试点、广东省科技厅"绿色低碳城区建设技术集成与示范区"重大专项科研示范单位。

Sino-Singapore Guangzhou Knowledge City is located in the east of Guangzhou. It is a key collaborative project and the third generation of its kind initiated by both China and Singapore. It is intended as a new engine for China's development of knowledge-based economy.

The main urban area of 32km^2, in the center of the Knowledge City is planned as a national-level regional innovation core under Sino-Singapore collaboration, an opening-up innovation area for service industry and S&T financial center in Pearl River Delta, an important base of innovation and entrepreneurship for excellent domestic and international talents, and a world-class livable ecological new town. The urban design intends to, through integration of space, economy, society, culture and ecology, assign special location value to the Knowledge City, create a leading intelligent and civilized place in Pearl River Delta, and eventually form an overall urban space pattern of mountain-river integration and compact cluster.

In terms of innovation, the planning: 1) explored methods and ways to turn urban design to regulatory detailed plan, upgraded urban design to a key link of planning management, and effectively elevated the urban space quality; 2) thoroughly practiced the technical method of "multi-plan in one" and ensured "implementation" of the plan; 3) established a means of green ecological index implementation with plan and guidelines as the carrier, and enabled the practice of "green ecology" concept. In terms of planning implementation, it 1) effectively instructed the development of over 150 works in the Knowledge City including municipal infrastructure, public service supporting facilities, and landscaping; 2) vigorously advanced the development of the Knowledge City as a green ecological urban area, and enabled it among the first pilot projects of national intelligent city by the MOHURD and key specialized scientific research models for "technology integration and demonstration area for green low-carbon urban area development" by Guangdong Provincial Department of Science and Technology.

项目地点:广州市黄埔区(原萝岗区)九龙镇
设计时间:2013.7-2013.12
建设时间:2014至今
用地规模:32km^2
建设规模:3050万m^2
合作单位:新加坡缔博(DP)建筑师事务所、广东省城乡规划设计研究院
曾获奖项:2015年第四届新加坡规划师学会奖 国际类最佳城市设计银质奖
2015年度全国优秀城乡规划设计奖(城市规划类)三等奖
2015年度广东省优秀城乡规划项目 一等奖

Location: Jiulong Town, Huangpu (former Luogang) District, Guangzhou
Design: 2013.7-2013.12
Construction: 2014 to date
Land area: 32km^2
GFA: 30,500,000m^2
Partner: PTE DP ARCHITECTS LTD
Guangdong Urban & Rural Planning and Design Institute
Awards: Silver Medal for Best International Urban Design, The Fourth Singapore Institute of Planners Award, 2015
The Third Prize of National Excellent Urban and Rural Planning and Design Award (Urban Planning), 2015
The First Prize of Excellent Urban and Rural Planning Projects of Guangdong Province, 2015

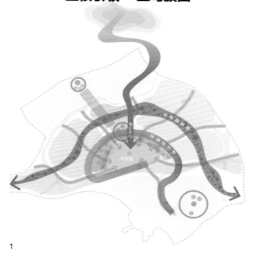

1

2

1 规划结构图
 Planning structure

2 城市设计总平面图
 Master plan of urban design

3 环湖核心区日景效果图
 Day view rendering of lakefront core area

4 九龙湖北岸夜景鸟瞰图
 Birds' eye view (night view) of the north bank of the Jiulong Lake

萝岗中心区·广州科学城城市设计及景观规划
Urban Design and Landscape Planning of Luogang Central Area · Guangzhou Science City

萝岗中心区、广州科学城是国家级高新技术产业园核心区、广州东部山水新城的公共服务核心。本次规划针对地区产城分离、职住失衡、人气活力不足等现状问题，把握地区生态本底优越、产业基础雄厚的特征，致力于营造持续的产业繁荣与城市活力，打造宜居宜业的新城典范。

规划布局与设计上特别强调"山水相融、产城融合、以人为本、多元活力"，提出五大规划策略：构建产城融合、职住平衡发展的格局；整合低效用地、挖掘释放土地价值潜力；突出"山、水、城"相融的景观特色；完善舒适、宜人、景观体验丰富的慢行系统；塑造地区门户形象地标，传承岭南地域文化特色。

探索大尺度城市空间规划方法，以景观规划为切入点，结合城市更新、存量规划、城市公共空间设计等手法，规划塑造了地区未来新的空间结构和功能布局，指导规划范围内多个规划管理单元控制性详细规划的编制，提出具体的导控条文，切实指导后续开发建设，促使本次规划的各项内容得到充分落实。

Luogang Central Area and Guangzhou Science City are the core of the national level hi-tech industrial park and the public service core of Guangzhou's Eastern Landscape New Town. By addressing the existing problems of industry-city separation, job-housing imbalance and insufficient vitality and taking advantage of the ecological strength and solid industrial foundation of the Area, the planning aims to foster continuous industrial prosperity and urban vitality and set a good example as a livable and entrepreneur-friendly new town.

The planning layout and design emphasizes the idea of "landscape integration, industry-city integration, people-orientation, diversified development". It proposes five planning strategies: achieving industry-city integration and job-housing balance; integrating low-efficient lands and explore their potential value; highlighting the landscape feature of "mountain, river and city integration"; establishing a comfortable, pleasant slow-traffic system with luxuriant landscape experience; shaping up the regional gateway image and inheriting Lingnan culture.

Starting from landscaping, the planning seeks to work upon large-scale urban spaces by means of city renewal, inventory planning, and urban public space design. On such basis, it shapes up the future spatial structure and functional layout of the Area to guide the regulatory detailed plans of several planning management units within the planning scope, and proposes concrete guiding and control provisions to instruct subsequent development and ensure thorough implementation of the planning.

项目地点：广州市黄埔区（原萝岗区）
设计时间：2014.07
规划面积：61.2km²
曾获奖项：2015年度全国优秀城乡规划设计奖（城市规划类）三等奖
2015年度广东省优秀城乡规划设计二等奖

Location: Huangpu District (former Luogang District), Guangzhou
Design: 2014.7
Size: 61.2km²
Award: The Third Prize of National Excellent Urban and Rural Planning and Design Award (urban planning), 2015
The Second Prize of Excellent Urban and Rural Planning and Design of Guangdong Province, 2015

1 土地利用规划图
  Land use plan
2 规划结构图
  Planning structure
3 萝岗中心区市民广场日景效果图
  Day view rendering of Luogang Central Area Citizen Square
4 市民广场岭南商业街
  Lingnan Shopping Street of the Citizen Square
5 香雪商业中心，塑造岭南风貌与现代商业商务中心的融合
  Xiangxue Commercial Center-an integration of Lingnan cityscape and modern commercial/business center

1

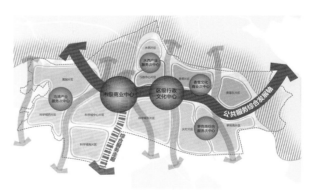

2

3

4

5

南沙新区蕉门河中心区中区和南区环境整治与景观提升设计总咨询
Consultancy for Environmental and Landscape Improvement of Jiaomen River Central Area (South/Middle Zone), Nansha

本项目是在南沙新区、自贸区"城市品质提升"规划建设管理体制机制创新的宏观背景下的首个试点项目,以景观总体设计为技术统筹平台,创新地探索跨部门整合、多层次衔接的规划实施咨询制度,具有重要的首创意义。

以国际化为标准,结合南沙水乡、冲积平原的特色,形成"积淀、千层"的设计概念,提出"都市生态绿洲,市民休闲中心"的设计定位。优化岸线,采用亲水平台、阶梯、微地形等,紧扣概念,既满足亲水性,也保证丰水期的使用。选用本土植物,营造滨海风情。结合景观总体设计元素,设计标示牌、城市家具等。引入海绵城市、生态廊道等前沿设计方法,采用清洁能源等成熟的低碳节能技术。

景观总体设计包括整体提升区、示范段及核心区三个层面的工作。整体提升区围绕蕉门河中心区"五个一"(一路、一河、一涌、一湖、一广场),构建地区整体景观框架。示范段包括市政绿园、商务花园、休闲健身公园三大节点。核心区打造市民广场、蕉门公园两条景观轴线,结合双桥形成品质化、艺术化的南沙标志性景观;通过增加水上运动等特色活动,加强中心区活力。统筹咨询阶段从设计到施工跟进了16项分项工程,涉及相关单位20余个,在总体层面衔接各项目,建设时间短,实施效果好。

The Project is the first pilot program in the macro-background of system and mechanism innovation for planning and development management of "upgrading city quality" by Nansha New Area and Free Trade Zone. It takes overall landscape design as a technical coordination platform to explore the planning implementation consulting system that features trans-department integration and multi-level connection. It is of great pioneer significance.

The design concept of "sedimentary deposits, thousand layers" and the design positioning of "urban eco-oasis, civic leisure center" are conceived as per international standards in view of Nansha's features as a water town and an alluvial plain. The shoreline is optimized with waterside platform, terrace and microtopography to echo the water-accessible theme and ensure wet season use. Indigenous plants are employed to foster coastal atmosphere. Signboards and urban furniture are designed in consideration of overall landscaping elements. Cutting-edge sponge city and eco-corridor approaches are introduced, while clean energy and other mature low-carbon energy conservation technologies are adopted.

Overall landscape design is implemented at three levels, i.e. overall upgrading area, demonstration area, core area. In the overall upgrading area, a regional overall landscape frame is established based on "Five Ones" (one road, one river, one riverway, one lake, one square) in Jiaomenhe Center. In the demonstration area, the landscaping covers three nodes: municipal green park, business garden, leisure & fitness park. In the core area, two landscape axis, i.e. civic square and Jiaomen Park, are provided to make quality, artistic landscape landmark in Nansha together with the double bridges; water sports and other featured activities are also added to enhance vitality. The integrated consulting stage covers 16 unit projects involving over 20 institutes from design to construction. It links up various projects on the overall level to ensure short construction yet satisfactory effect.

项目地点:广州市南沙区蕉门河中心区
设计时间:2015.1-2015.7
建设时间:2015.7
规划面积:规划范围内约5km²(整体提升区),其中示范段约1km²,核心区约19公顷
合作单位:艾奕康环境规划设计(上海)有限公司广州分公司(AECOM)
曾获奖项:2014-2015年度广东省优秀工程咨询成果奖 一等奖

Location: Jiaomenhe Central Area, Nansha District, Guangzhou
Design: 2015.1-2015.7
Construction: 2015.7
Size: about 5km² (overall upgrading area) within the planning scope, including about 1km² of demonstration area and 19ha of core area
Partner: AECOM Guangzhou Office
Award: First Prize of Excellent Engineering Consultation Accomplishment Award of Guangdong Province, 2014-2015

1 精细化、一体化设计的开放广场,与四大馆、双桥景观和谐统一
Open square with refined and integrated design in harmony with the four cultural facilities and two bridges

2 以开放公共空间承载多元活动,塑造广场地标,设置立体的视觉观赏点
The pubic open spaces are created to accommmdate the various events while providing landmark and multi-level viewing points

3 强化视线和交通的联系,塑造怡人丰富的亲水活动空间
The connection between sightline and traffic is strengthened, while pleasant and diversified water-accessible spaces are created

4 蕉门河一河两岸、双桥景观实景
View of Jiaomen River with two bridges

3

4

广州番禺万博商务区地下空间控制性详细规划
Regulatory Detailed Plan of Underground Space of Wanbo CBD, Guangzhou

广州番禺万博商务区地下空间是广州新一轮地下空间开发试点工程，是广州市贯彻落实新型城市化发展的重要体现。项目作为广州首个地下空间控制性详细规划，填补了地下空间控规的空白。首次提出并应用"地下空间缓冲区"，实现地下空间连片统一规划、无缝衔接。

项目贯彻"多专统筹，上下一体"及"以人为本，经济可行"的规划理念。通过将规划、建筑、市政、景观、机电设备等多个设计专业进行统筹、宏观协调，利用规划引导各功能分区之间横向和纵向高效衔接，形成上下一体的立体组织，增强公共交通的辐射能力，提高区域运营效率和服务水平。设计尊重项目地形特点，尽量减少不必要的开挖，考虑人的活动和自然资源的联系，引入下沉绿地、公共下沉广场、构筑极具岭南特色的半地下地下空间，制定怡人的地下空间尺度控制，在满足空间使用的安全、便利、舒适的情况下，达到节能、经济的效果，并强调规划的科学性和可实施性。

万博地下空间将以地铁7号线万博中心站为核心，打造融合交通、商业、公共服务、停车、绿色市政的复合型地下综合体，地下共分四层开发，地下总开发建筑面积约为180万平方米。核心区目前8个地块实现区域同步开挖、分步快速建设。本项目对地下空间控规成果、控制指标的探索，首创"地下空间缓冲区"的实践经验，成功推广至省、市多个重点地区地下空间规划。同时，项目作为重点案例纳入住建部《特大城市重点地区地下空间规划技术研究》课题研究。

The Project is a new round of Guangzhou's pilot development of underground space and an important showcase of the city's implementation of new urbanized development. Its regulatory detailed plan is the first for underground space in Guangzhou. With the innovative concept of "underground space buffer area", it achieves integrated planning and seamless connection of underground spaces.

The planning concept is "multi-disciplinary integration, vertical organization" and "people orientation, cost effectiveness". By integrating planning, architecture, utilities, landscape and MEP disciplines and guiding effective horizontal/vertical connections between various functional zones via planning, the Project establishes a vertical system that enhances the coverage of public transport and regional operation efficiency/service level. The design respects the site terrain by minimizing unnecessary excavation and gives consideration to human-nature connection by introducing sunken green space, public sunken square, and semi-basement with Lingnan characteristics and creating pleasant basement scale. While ensuring safe, convenient and comfortable use of space, the design also achieves energy efficiency and cost effectiveness while highlighting the scientificity and implementability of the planning.

With Wanbo Center of Metro Line 7 as the core, the design intends to create a mixed-use underground complex integrating traffic, commercial, public service, parking and green municipal facilities. The underground space will include four floors totaling about 1,800,000m². At present, eight plots in the core area are being excavated simultaneously and undergoing the phased construction. Based on the project's exploration on the regulatory planning and indices of the underground space, the "buffer zone of underground space" is initiated as a practical experience and has been applied for the planning of underground spaces in key areas of the city and the province. Meanwhile, the project is included into MOHURD's <Research on Underground Space Planning Technology in Key Areas of Mega Cities>.

项目地点：广州市番禺区万博中心
设计时间：2012
建设时间：2012年至今
建筑面积：规划范围1.5km²，规划地下空间总规模180万m²（其中计容总规模37万m²）
建筑层数：最大开发深度地下4层
曾获奖项：2015年度广东省优秀城乡规划设计三等奖

Location: Wanbo CBD, Panyu District, Guangzhou
Design: 2012
Construction: 2012 to date
GFA: Planning scope – 1.5sq.km., Planned underground space -1,800,000m² (FAR-counted GFA: 370,000m²)
Floors: Max. 4 floors underground
Award: The Third Prize of Excellent Urban and Rural Planning and Design of Guangdong Province, 2015

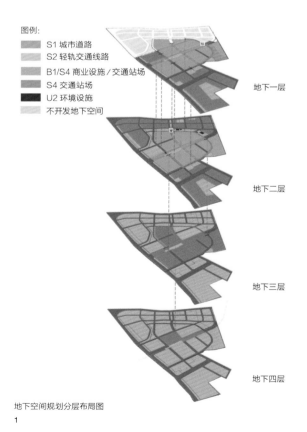

地下空间规划分层布局图

1. 地铁核心九大地块无缝衔接，紧凑集约利用轨道站点周边地下空间资源
Nine major plots at the core of the metro network are seamlessly connected to realize the compact and intensive use of the underground spaces around the metro stations

2. 利用现状地形以退台方式构造半地下特色骑楼街
Semi-underground arcade street is created via setbacks on existing terrains

3. 整合、连接各地块地下空间建筑，建设融合交通、商业、公共服务、停车、绿色市政的复合型地下综合体。
3. The underground developments of various plots are integrated and connected to create a mixed-use underground complex that integrates transport, retail, public services, parking and green utilities

4. 利用下沉广场作为人流疏散及地下空间的出入口，结合景观设置提升地下空间的趣味与生态性
Sunken plazas are provided as access of people evacuation and underground spaces. The landscaping is integrated to create a more attractive and ecological underground space

5. 使用楼梯、自动扶梯及垂直电梯解决地上地下人流的竖向联系
Stairs, escalators and elevators are provided to realize the vertical connection between the pedestrians below and above the grade

6. 利用汉溪大道中央绿化隔离带设置下沉式公共绿地，为地下室商业提供室外阳光和自然通风
Sunken public green is provided based on the Han Xi Da Dao green isolation belt, allowing for daylight and natural ventilation in the underground spaces

2

3

4

5

6

市政桥道

1952年建院至今，GDAD一直从事市政道桥及隧道工程设计，2002年将市政专业团队定名为市政工程设计院。20世纪80年代以前参加了广东省内佛山、茂名、湛江、汕头、韶关、肇庆等地市的多项市政道桥项目，其中江门蓬江大桥为广东省内首座无支架悬索吊装施工的箱形拱桥。80年代至2000年期间设计了国内首条城市高架桥六二三高架桥、惠州东江大桥、开平大桥、江门西江潮连大桥、英德北江浈阳大桥等市政道桥工程，其中六二三高架桥获得了国家优秀设计金质奖。

2000年以来市政工程设计院蓬勃发展，不断开拓新领域，从广州内环路开始我院陆续参与了广州永九快速路、番禺南大干线等多项城市快速路的设计，其中广州内环路人民桥北一南田路段及镇安路立交工程获国家优秀工程银质奖；番禺南大干线全长30.32公里，全线共设立交24处，本项目线路长，立交类型多，现场条件复杂，协调难度大，此项目设计提高了GDAD超长超复杂城市快速路和互通立交领域的设计水平。GDAD还负责了广州铁路新客站地区、广州白云新城、金沙洲居住新城、广州南沙新区明珠湾区起步区、郑州航空港区等多个不同类型区域路网的设计，对区域交通组织设计具有丰富的经验，其中广州铁路新客站地区市政及附属工程项目智能交通组织历经春运、黄金周等交通高峰考验，并获得全国优秀工程设计一等奖。随着城市用地的限制和景观要求的提高，城市隧道和地下空间开发日益增多，近年来GDAD完成了二十多条城市隧道设计，其中华南路石陂隧道为广州市区最长的双洞穿山隧道；通过广州珠江新城市政交通项目、万博地下空间、广州国际金融城起步区地下空间等项目GDAD在地下道路系统设计方面处于国内领先水平。在传统梁式桥、拱桥设计的基础上GDAD在特大斜拉桥设计领域取得新发展，江湾大桥为33米+102米+183米钢混组合独塔斜拉桥，主塔采用拱塔结构造型，主梁宽度达到44.5米，为同类桥型中最宽的桥梁，沥桂大桥为160米+120米独塔斜拉桥，主塔采用异形"A"字形塔。随着城市边界的不断拓展，城郊区域的公路必须改造为市政道路，GDAD在广州科学城天麓南路扩建工程、105国道、355省道城区段市政改造工程、江门金瓯路等旧路改造项目中取得了很好的效果，升级改造后由过境公路转变为功能齐全、安全顺畅、景观优美的市政样板路。

GDAD市政工程设计院人才梯队科学，技术力量雄厚，设计经验丰富。团队在城市高快速路、大型互通立交、城市区域路网、特大桥梁、特长隧道、地下空间和旧路改造等领域设计达到国内先进水平，设计业绩突出，得到广泛好评，先后累计获得全国或省级优秀工程设计奖33项。同时坚持"诚信第一，设计第一"，充分利用人才、技术、科研、创新和品牌的综合优势，为广大客户提供高效优质的服务，共同设计未来，成就梦想。

Bridges and Roads

Since its establishment in 1952, GDAD has been engaged in the engineering design for municipal road, bridge and tunnel. In 2002, the municipal team was named as Municipal Engineering Design Institute. Before the 1980s, we participated in a number of municipal road and bridge projects in Foshan, Maoming, Zhanjiang, Shantou, Shaoguan, Zhaoqing and other cities in Guangdong Province, among which, Jiangmen Pengjiang Bridge is the first box arch bridge of suspension cable-hoisting construction without supporting frame. During the period from 1980s to 2000, we designed various municipal bridges, including the first urban viaduct 623 Viaduct, Huizhou Dongjiang Bridge, Kaiping Bridge, Jiangmen Xijiang Chaolian Bridge, Yingde Beijiang and Zhenyang Bridge, etc., and among which 623 Viaduct won the Gold Prize of National Excellent Engineering Design Award.

Since 2000, Municipal Engineering Design Institute has witnessed rapid development and constantly expanded service scope. Beginning with Guangzhou Inner Ring Road, we participated in the design of Guangzhou Yongjiu Expressway, Panyu South Main Line, and other urban expressways. The interchange engineering from Renmin Bridge North to Nan Tian Lu and Zhen An Lu won the Silver Prize of National Excellent Engineering Design Award; the 30.32km Panyu South Main Line with 24 interchanges, involves difficult coordination due to the extraordinary length, multiple types of interchanges and complicated working environment. With this project, we greatly improved our design expertise in extra-long and complicated urban expressways and interchanges.

We are highly experienced at regional transportation organization after designing various types of regional road networks, including New Guangzhou Railway Passenger Station, Guangzhou Baiyun New Town, Jinshazhou New Residential District, Pearl Bay District Kick-off Area of Nansha New District, and Zhengzhou Airport Zone, etc. The intelligent transport organization of municipal and auxiliary project for New Guangzhou Railway Passenger Station successfully withstood the traffic peaks during the Spring Festival and the Golden Week, and won the First Prize of National Excellent Engineering Design Award. In response to the more stringent requirements on urban land and landscape, more urban tunnels and underground space are developed.

In recent years, we have completed the design of over 20 urban tunnels, including the longest two-track mountain tunnel, Hua Nan Lu Shibei Tunnel. Now we are a leader in design of underground road system in China thanks to our project practices with the Pearl River New Town's Municipal Transport Project, Wanbo Underground Space, and Guangzhou International Financial City Kick-off Area Underground Space, etc..

On top of conventional beam bridge and arch bridge, we have made fresh progress in the design for super large cable-stayed bridge. Jiangwan Bridge, the 33m+102m+183m single-pylon cable-stayed bridge of steel-concrete combination, features arch pylon structure and 44.5m-wide girder, the widest among the same kind. Ligui Bridge, 160m+120m single-pylon cable-stayed bridge, adopts unconventional "A"-line pylon. Along with the urban expansion, the suburban roads must be transformed into municipal roads. So far our efforts have achieved satisfactory result with Tian Lu Nan Lu extension project in Guangzhou Science City, urban section transformation project of No.105 National Highway and No.335 Provincial Road, and Jin Ou Lu renovation project in Jiangmen. These transit roads have been upgraded and transformed into functional and safe exemplary municipal roads with smooth traffic and attractive landscaping.

The Municipal Engineering Design Institute is well-established with talents, expertise and project experiences. As a leader in engineering of urban expressways, large interchanges, regional urban road network, super large bridges, extra-long tunnels, underground space and old road transformation in China, we have been extensively recognized for our remarkable portfolio and received 33 national or provincial excellent engineering design awards. Valuing professional ethnics and quality design, we will take full advantage of our comprehensive strength in talent, technology, research, innovation and brand to offer our clients efficient and quality services for a win-win result.

广州铁路新客站地区市政及附属工程
Municipal and Ancillary Works in New Guangzhou Railway Passenger Station Area

广州南站（广州铁路新客站）是全国首条开通的高铁南端终点站，是广州乃至珠三角地区铁路客运南大门。以广州南站为核心的广州铁路新客站地区将打造成华南枢纽、活力新城。区域市政道路及附属工程建设，是广东省基础设施建设重点项目。

本项目为广州南站配套市政建设项目，通过区域交通组织，辅以区域智能交通诱导系统，疏导进出广州南站交通，历经春运、黄金周等交通高峰考验，卓有成效；在极其有限的工期内，通过设计优化，采用先进技术及施工组织，仅用6个月时间，完成面积高达126.5万平方米道路软基处理、1.42公里长单向双车道高架匝道桥梁、下沉式立交隧道4座、下沉式道路2座，按期确保广州南站的通车运营；采用新型材料如仿花岗岩石混凝土人行地砖、LED护栏照明系统等以较少的投资，带来较高的道路景观效果。

本项目总投资约45亿元，建设城市主干路、城市次干路共26公里，其中城市I级主干路12公里，规划标准宽度为40~80米，道路总面积约为120万平方米；其中包含下沉式立交隧道4座，总长2.2公里；下沉式道路2座，总长650米；高架匝道桥6座，总长2.9公里；跨江中桥一座，长96米；人行地下通道11座，人行天桥4座，景观排水明渠5公里。

Guangzhou South Railway Station (New Guangzhou Railway Passenger Station, the Station) serves as the south terminal station of the first operating high-speed railway in China and the south portal to railway passenger transport in Guangzhou and even the Pearl River Delta. The Station Area, with the Station at the core, is planned as a transportation hub and dynamic new town in South China. Development of municipal roads and ancillary works in this Area represents a key infrastructure project of Guangdong Province.

As the supporting municipal works of the Station, the Project aims to distribute the traffic to/from the Station through local traffic organization and intelligent traffic guidance system, proven to operate effectively during the peak hours such as the Spring Festival and the golden weeks. Within a very tight project schedule, it took only 6 months to complete the soft foundation treatment for 1.265 million square meters of road, 1.42km-long single-way double-lane elevated ramp bridge, 4 underpass tunnels and 2 sunken roads, ensuring the scheduled operation of the Station. In addition, new materials such as granite-like concrete footpath tiles and LED guardrail lighting system are adopted to achieve attractive road landscape with less investment.

With a total investment of about RMB4.5 billion, the Project develops 26km-long urban primary trunk roads and secondary trunk roads, including 12km-long Grade I urban primary trunk road with a planned typical width of 40-80m and a gross road area of about 1.2 million m^2, 4 underpass tunnels with a total length of 2.2km, 2 sunken roads with a total length of 650m, 6 elevated ramp bridges with a total length of 2.9km, one river-crossing bridge with a length of 96m, 11 underground pedestrian passages, 4 pedestrian bridges and 5km-long landscaped open ditch.

项目地点：广州市番禺区
设计时间：2008.11
建设时间：2009.5
工程规模：城市区域路网设计，城市主次干道共26km，其中包含隧道4座，总长2.8km；高架桥梁7座，总长3.0km；人行地下通道11座，人行天桥4座，景观排水明渠5km
曾获奖项：2013年度全国优秀工程勘察设计行业奖市政公用工程道桥一等奖
2013年度广东省优秀工程勘察设计奖一等奖

Location: Panyu District, Guangzhou
Design: 2008.11
Construction: 2009.5
Size: urban road network design; 26km-long urban primary truck roads and secondary trunk roads, including 4 tunnels with a total length of 2.8km, 7 viaducts with a total length of 3.0km, 11 underground pedestrian passages, 4 pedestrian overpasses and 5km-long landscaped open ditch
Awards: The First Prize of National Excellent Engineering Exploratio and Design Award (Road/ Bridge of Municipal Public Works), 2013
The First Prize of Excellent Engineering Design of Guangdong Province, 2013

1

1 总平面图
  Master Plan
2 通衢大道
  Thoroughfare
3 轻巧的钢筋混凝土叠合梁与新客站立面完美融合
  Light steel concrete superposed beams integrates perfectly with the facades of the station building
4 广州门户、路网通达
  Gateway of Guangzhou enjoys well-established road network
5 立面图
  Elevation
6 下沉式道路确保中轴广场景观和人行交通的便捷
  Sunken roads ensures the landscaping and easy pedestrian traffic of the Central Axis Square
7 定向隧道畅通,立面石材装饰美观
  Directional tunnel ensures smooth traffic while stone façade completes attractive appearance

番禺南大干线
Panyu South Main Line

番禺南大干线是省级重点工程，位于番禺北部，路线总体呈东西走向，全长27.2公里，被誉为番禺新城交通"大动脉"。全线起于番禺区石壁街东新高速（广州南站商务区），止于石楼镇莲花大道，是联系广州南站—番禺新城组团及大学城组团的交通干线。未来将形成佛山—番禺—南沙的快速通道，是连接广佛莞三地的重点交通基础设施建设项目之一。

本项目规划红线宽度80～90米，主辅分离，主线双向八车道，设计行车速度为80公里/小时，两侧各设单向双车道辅道，辅道设计行车车速为40公里/小时。与沿线城市主干道等级及以上道路立交，与次干路及以下道路采用右进右出。共设置立交24处，包含互通式立交5座，下穿隧道8座，跨线桥11座。项目建设内容涵盖包含道路、桥梁、隧道、给排水、照明、交通、绿化、电力管廊管沟等附属工程。

番禺南大干线定位为城市快速路，沿线与6条高快速路、18条主干道相交，合理选择立交结构形式以及结合主辅出入口设置，以满足区域交通转换功能。除此之外，道路沿线跨越番禺区人口密集的石壁街、大石街、南村镇，建设用地条件复杂，道路设计既需要改线避让众多的控制因素，还需要进行断面优化调整，在满足交通需求的同时兼顾施工可行性以及工程规模的控制。本项目定位于城市快速路的同时，兼顾服务于周边地块的功能的综合设计，以及立交节点方案选型等成功设计，为类似工程起到一定的参考和借鉴作用。

Running from the west to the east in the north of Panyu District with a total length of 27.2 km, Panyu South Main Line is a provincial key project praised as the main traffic artery in Panyu New Town. The whole line serves as a trunk road to link Guangzhou South Railway Station, Panyu New Town and GZHEMC. It starts at Dongxin Expressway, Shi Bi Jie, Panyu District (Guangzhou South Railway Station Business District) and ends at Lian Hua Da Dao in Shilou Town. As part of Foshan – Panyu – Nansha Fast Track in future, Panyu South Main Line represents a key traffic infrastructure project which connects the three cities of Guangzhou, Foshan and Dongguan.

With a planned width of 80-90m, the Project consists of the main line and auxiliary roads which are separated from each other. The main line comprises two-way eight lanes at the designed travel speed of 80km/h, while single-way two lanes as the auxiliary road are provided on both sides at the designed travel speed of 40km/h. Panyu South Main Line intersects with the highways at the level of the urban primary trunk road and above through grade separation, and the highways at the level of the secondary trunk road and below through grade crossing, with the entry and exit provided on the right side. The Project has 24 overpasses (including 5 interchanges) in total, 8 underpass tunnels and 11 flyovers. The construction contents cover such ancillary works as roads, bridges, tunnels, water supply/drainage, lighting, traffic, greening, and electrical pipe corridor/ditch.

Defined as an urban expressway, Panyu South Main Line intersects with 6 expressways and 18 primary trunk roads. Reasonable grade separations are employed to accommodate the regional traffic transfer together with the main and auxiliary accesses. In addition, as the Line passes through the densely populated Shibi Sub-district, Dashi Sub-district and Nancun Town where the construction conditions are complicated, the road is rerouted to evade a number of constraints and the road sections refined properly to allow for the construction feasibility and construction size control while meeting the traffic demands. The project sets an example for similar projects with its comprehensive function as an urban expressway that also supports the surrounding plots, and its grade separation solution.

项目地点：广州市番禺区
设计时间：2014.3
工程规模：城市快速路，道路长度约27.2km，立交24座，包含互通式立交6座，下穿隧道7座，跨线桥11座

Location: Panyu District, Guangzhou
Design: 2014.3
Size: urban expressway, with a road length of about 27.2km, 24 overpasses (including 6 interchanges), 7 underpass tunnels and 11 flyovers.

1   新化快速路互通立交全景
    A panorama of Interchange on Xinhua Expressway

2   总平面图
    Site Plan

3   标准路段主辅布置合理
    Reasonable layout of the Main line and Auxiliary Roads at the Typical Section

4   南大主线下穿，地面与新造路十字灯控
    South Main Line underpasses with traffic light control at the ground crossing with Xin Zao Lu

5   主线快速，辅道与沿线地块街接，通过立交实现交通转换
    With express traffic at the Main Line, the auxiliary road connects with plots along the Line and realizes traffic transfer via grade separation

6   南大主线上跨，地面与金光西路十字灯控
    South Main Line overpasses with traffic light control at the ground crossing with Jin Guang Xi Lu

7   双向八车道隧道洞口及敞口段近景
    A close view of the opening of a two-way eight-lane tunnel

1

2

3

4

5

6

7

Bridges and Roads 市政桥道 **277**

华南路三期工程
Huanan Expressway (Phase III)

广州市华南快速路三期工程位于白云区以北，呈东西向横贯全区，西起广和大桥联结佛山里水，东接华南快速路二期的春岗互通，联系了广清高速公路、广花高速公路、机场高速以及为数众多的城市道路。华南快速路三期工程的建设完善了广州市路网功能，并对促进地区经济均衡发展取到十分重要的作用。

广州市华南快速路三期工程C标段为城市快速路，计算行车速度80公里/小时，全长4.29公里，双向6车道。其中设置主线大中桥1327米/7座，穿山隧道长1020米，互通立交2处。桥隧等构造物占整个工程的55.4%。此工程获全国优秀勘察设计二等奖，广东省优秀勘察设计一等奖。

广州市华南路三期工程C标段地形地质条件非常复杂，瘦狗岭地质大断裂带通过本工程，断裂带裂隙非常发育，且道路沿线地势起伏大，相对高差达到200余米，同时还经过一处深约70米的采石坑，穿山隧道上方存在白云山石陂水库。除此之外，道路沿线有南湖水厂加压站，帝湖山庄，南湖煤气站等多处控制点需要避让。道路设计既需要避开众多的控制因素，还需要考虑路堑高边坡及深填方的安全性与经济性、白云山石陂水库对隧道施工的影响、工程规模的控制等。我单位凭借丰富的工程经验，经过精细化的设计，在满足行车舒适性的同时，确保了工程安全，且做到了工程规模最省，受到业主好评。其中的路堑高边坡、深填方路基、经过裂隙发育带的穿山隧道等成功设计，为类似工程起到一定的参考和借鉴作用。

项目地点：广州市白云区
设计时间：2003.7
建设时间：2005.11
工程规模：城市快速路，道路长度4.3公里，路基宽度28.5m，双向六车道，计算行车速度80km/h。其中设主线大中桥1327m/7座，双孔长隧道1020m/座，互通式立体交叉2座
曾获奖项：2011年度全国优秀工程勘察设计行业奖市政公用工程二等奖
2011年度广东省优秀工程勘察设计奖一等奖

Location: Baiyun District, Guangzhou
Design: 2003.7
Construction: 2005.11
Size: Two-way six-lane urban expressway with a length of 4.3km and a roadbed width of 28.5m and a calculated travel speed of 80Km/h; including seven large and medium-sized main-line bridges totaling 1,327m, one 1,020m – long double-hole tunnel and two interchanges.
Awards: The Second Prize of National Excellent Engineering Exploration and Design (Municipal Public Works), 2011
The First Prize of Excellent Engineering Design of Guangdong Province, 2011

The project runs from Guanghe Bridge on the west to Chungang Interchange of Huanan Expressway (Phase II) on the east across the north of the whole Baiyun District. It connects Lishui, Foshan, and links up Guangzhou – Qingyuan Expressway, Guangzhou – Huadu Expressway, Airport Expressway, and a number of urban roads. The project has improved the city road network and greatly contributed to the balanced development of the local economy.

Bid Section C of Huanan Expressway (Phase III) is a 4.29km-long two-way six-lane urban expressway with a calculated travel speed of 80Km/h. It has seven large and medium-sized main-line bridges totaling 1,327m, one 1,020m–long mountain tunnel and two interchanges. Such structures as bridges and tunnel account for 55.4% of the whole Project. The Project has won the Second Prize of National Excellent Engineering Exploration and Design of China, and the First Prize of Excellent Engineering Exploration and Design of Guangdong Province.

The project is plagued by highly complicated geological conditions. Fully developed large fault zone of Shougou Hills passes through the site. The terrain along the road varies greatly, with a relative height difference of over 200m. There is a 70m-deep stone pit on the route, and Shibei Reservoir of Baiyun Mountain above the mountain tunnel. In addition, the Project also needs to avoid multiple control points such as the Pressure Station of Nanhu Water Plant, Dihu Resort Compound, and Nanhu Gas Station, and consider the safety and cost effectiveness of cutting high slope and deep earth filling, impact of Shibei Reservoir of Baiyun Mountain on the tunnel construction, and engineering size control. With considerable project experiences and meticulous design, we have ensured both the travel comfort and engineering safety, and minimized the engineering size, winning an acclaim from the Client. The successful design practices such as cutting high slope, deep earth filled roadbed, and mountain tunnel through the fracture development zone all sets examples for similar projects.

2

1 总平面图
   Site plan

2 同和立交匝道
   Tonghe overpass ramp

3 帆船造型的隧道洞口
   Sailboat-shaped tunnel opening

4 绿色生态高边坡景观优美
   Attractive green and ecological high slopes

5 沙太大桥与石陂隧道衔接远景
   Distant view of connection between Shatai Bridge and Shibei Tunnel

6 单向三车道隧道内景
   Inner view of one-way three-lane tunnel

7 同和跨线桥
   Tonghe Flyover

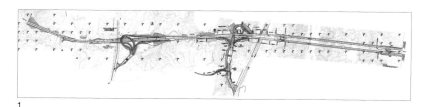

1

3

4

5

6

7

Bridges and Roads 市政桥道

市政给排水

自1952年建院以来，GDAD一直致力于市政给排水及固废处理工程设计及研究工作。设计及研究领域包括：污水处理工程、供水工程、黑臭水体、海绵城市、综合管廊、环境卫生、固废处理等专业，是广东省水处理及固废处理的主要设计力量。

GDAD市政给排水及固废处理专业人才梯队科学，技术力量雄厚，设计经验丰富，达到国内先进水平。现有专业技术人员247名，其中教授级高级工程师6名，高级工程师86名，中级工程师56名，国家注册工程师28名。GDAD不断吸收和培养业界创新人才，精心设计，提高创新能力和服务意识，始终掌握行业的前沿技术和科学的设计手段，关注行业建设热点，努力解决本行业的难点问题，为用户提供优质的设计和细致的服务，设计作品遍布广东省各地及国内其他城市，并得到广泛好评。

在设计创优方面，近年来先后获得全国优秀设计、广东省优秀设计等奖项四十多项，其中广州大坦沙污水处理厂（三期）系统工程、广州兴丰生活垃圾卫生填埋场工程获得全国优秀工程设计铜奖、广州新塘水厂技术改造工程获得全国优秀工程勘察设计一等奖。获得省级优秀设计奖项的还有：广州市西江引水工程、广州市花都区新华污水处理厂（一、二期）提标改造工程、广州市江高-石井污水处理系统工程、广州石井污水处理系统工程、广州市废弃物安全处置中心一期工程等项目。在科技创新方面，依托"广东省现代建筑设计工程技术研究中心"和"广东省水环境与生态工程技术研究中心"两个省级科研中心，在原水长距离输送技术、黑臭水体处理工艺等方面先后完成了十多项省级以上科研创新成果，并获得发明专利、实用新型专利及科技进步奖多项。

GDAD是全国最早开展公用设施特许经营招商咨询的单位之一，咨询团队擅长新时期PPP项目的需求和绩效研究，已成为全国多个省市PPP项目咨询服务的入库单位。

近年来，城市地下综合管廊、海绵城市建设和黑臭水体整治成为国家的建设热点，GDAD充分发挥了科技和人才优势，为省、市政府出谋划策，先后编制了《广东省城市基础设施建设"十三五"规划》、《广东省城市地下综合管廊实施总体方案》、《广州市市政综合管廊工程施工及验收规范》，为我省的综合管廊、海绵城市建设和黑臭水体整治提供了政策和技术保障。同时也承担了广州国际金融城起步区地下综合管廊、万博商务区地下综合管廊、琶洲西区（互联网创新集聚区）综合管廊、广州南沙新区灵山岛尖综合管廊等十多项工程的设计；肇庆新区城市地下综合管廊建设专项规划编制等工作；跨越珠海、澳门二地的鸭涌河整治工程、广州及各地市的多项黑臭水体整治工程设计工作也在同步进行中。

GDAD将继续秉承"守正鼎新，营造臻品"的核心价值观，发扬"绘雅方寸，筑梦千里"的企业精神，充分利用人才、技术、科研、创新和品牌的综合优势，为广大客户提供高效优质的服务，共同设计未来，成就梦想。

Municipal Water Supply & Drainage

Since its founding in 1952, GDAD has been a main design force in water and solid waste treatment of Guangdong Province, committed to the design and research of municipal water supply/drainage, and solid waste treatment projects, covering such fields as sewage treatment, water supply, black and odorous water bodies, sponge city, utility tunnel and environmental health and solid waste treatment.

As a leader in the field of municipal water supply, drainage and solid waste treatment in China, our team are well-established with talents, expertise and project experiences. Currently the team is staffed with 247 professionals, including 6 professorial senior engineers, 86 senior engineers, 56 intermediate engineers and 28 registered engineers. On one hand, we have been focusing on building the innovation talent team and improving our design expertise, innovation capability and service awareness; on the other hand, we always keep an eye on the cutting edge technologies and design approaches as well as the latest development in the industry, striving to explore solution to the key problems of the industry. With design projects all over Guangdong Province and other Chinese cities, we have been extensively acclaimed for quality design and all-around services rendered to the clients.

In terms of honors and awards, we received over 40 national and provincial excellent engineering design awards in recent years. Datansha Sewage Treatment Plant (Phase III) and Xingfeng Domestic Waste Sanitary Landfill in Guangzhou won the Bronze Prize of National Excellent Engineering Design Award, and Technical Improvement of Guangzhou Xintang Water Plant won the First Prize of National Excellent Engineering Exploration and Design Award. Projects winning the provincial excellent engineering design awards include Guangzhou Xijiang River Diversion, Upgrading and Improvement of Xinhua Sewage Treatment Plant (Phase I, Phase II), Huadu District, Guangzhou, Guangzhou Jianggao – Shijing Sewage Treatment System, Guangzhou Shijing Sewage Treatment System, and Guangzhou Safe Waste Disposal Center (Phase I). In terms of technological innovation, we have completed more than ten provincial scientific research innovations on long-distance raw water transport technology and black and odorous water body treatment technology, and won multiple patents for invention and utility models and awards for technological advancement, with the support of two provincial scientific research centers, namely Guangdong Modern Architectural Design and Engineering Technology Research Center and Guangdong Water Environment and Ecological Engineering Technology Research Center.

GDAD is among China's first institutions that offer consultancy on attracting investment in utility franchise. As our consultation team excels at demand and performance research of PPP projects, we have been included in the PPP project consultancy banks of multiple provinces and cities in China.

In recent years, China has been focusing on the construction of urban underground utility tunnel and sponge city, and black and odorous water body treatment. Giving full play to its technological and talent advantage, GDAD has acted as advisor to the provincial and municipal governments and developed the *Thirteenth Five-year Plan for Urban Infrastructure of Guangdong Province*, *Overall Implementation Scheme of Urban Underground Utility Tunnel of Guangdong Province*, and *Code for Construction and Acceptance of Municipal Utility Tunnel of Guangzhou City*, giving policy and technical support to construction of utility tunnel and sponge city construction and black and odorous water body treatment. In addition, we have also undertaken the design for over then projects such as the underground utility tunnel of Kick-off Zone of Guangzhou International Financial City, the underground utility tunnel of Wanbo Business District, utility tunnel in Pazhou West Area (Internet innovation concentration area) and utility tunnel of Lingshan Island Tip, Nansha New Area, Guangzhou, and developed specialized construction plans of the urban underground utility tunnel of Zhaoqing New Area. The design for the improvement of Yachong River over Zhuhai and Macau and multiple improvements of black and odorous water bodies in Guangzhou and other cities is also underway at the same time.

Carrying on the core values of integrity, responsibility, innovation and development and the corporate spirit to design and build for dreams, we will take full advantage of our comprehensive strength in talent, technology, research, innovation and brand to provide our clients efficient and quality services for a win-win result.

广州市西江引水工程
Xijiang River Water Diversion Project, Guangzhou

本项目是广州市迎接亚运会、保障供水水源安全、提高供水水质的重点工程，对于满足和促进广州市城市化建设的要求，保障居民身体健康有十分重大的意义。

工程设计规模350万立方米／天。地点位于广州市与佛山市境内，自佛山市三水区西江思贤滘下陈村附近的西江取水，经下陈取水泵站增压后，通过2×DN3600 管道输至鸦岗配水泵站，原水通过鸦岗配水泵站分配提升，通过管道输送至广州市西部白云区的江村水厂（40万立方米／天）石门水厂（80万立方米／天）和西村水厂（100万立方米／天），工程总投资约90.0亿元人民币。

输水管道干线全长约47.09公里，输水主干线管径2×DN3600 在穿越大型交通枢纽且采用盾构穿越时，其管径采用1×DN4800输水管道主管材为预应力钢筒混凝土管（PCCP），管线穿越障碍采用钢管（SP）。穿越北江时采用沉管方式，一次性沉管长度为540米，管径为DN3600。

As a key project of Guangzhou to support the 2010 Asian Games, ensure the security of water supply and improve the quality of drinking water, the project is of great significance for meeting and promoting the city's urbanization process and safeguarding the health of the citizens.

Designed with a capacity of 3.5 million m³/d, the project is located in the cities of Guangzhou and Foshan. The water is taken from Xijiang River near Xiachen Village upon Sixianjiao River, a tributary of Xijiang River located in Sanshui District, Foshan, and sent through 2 × DN3600 pipeline to Yagang Distribution Pump Station after pressurization in Xiachen Village Intake Pumping Station. The raw water is then sent through pipelines to Jiangcun Water Plant (400,000 m³/d), Shimen Water Plant (800,000 m³/d) and Xicun Water Plant (1million m³/d), all located in Baiyun District in the west of Guangzhou, after distribution and pressurization in Yagang Station. The total investment of the Project is about RMB 9 billion.

The water pipeline extends 47.09km and the diameter of the main water pipe is 2 × DN3600. When penetrating large-scale transportation hubs where shield tunneling method is adopted, prestressed concrete cylinder pipe (PCCP) with diameter of 1 × DN4800 is used as main pipe while steel pipe (SP) is used where the pipeline penetrates obstacles. Immersed tube method is adopted when the pipeline runs through the Beijiang River, with the one-time immersed tube measuring 540 m long and 2 × DN3600 in diameter.

项目地点：广州市与佛山市境内
设计时间：2004.6
建设时间：2009.7
建筑面积：设计规模350万m³/d
合作单位：中国市政工程华北设计研究总院、中水珠江规划勘测设计有限公司
曾获奖项：2015年度广东省优秀工程勘察设计奖二等奖

Location: Guangzhou and Foshan
Design: 2004.6
Construction: 2009.7
GFA: 3.5 million m³/d
Partner: North China Municipal Engineering Design and Research Institute (NCME)
China Water Resources Pearl River Planning Surveying & Designing Co., Ltd
Award: The Second Prize of Excellent Engineering Design Award of Guangdong Province, 2015

1 总平面图
Site plan

2 DN3600钢管焊接施工
DN3600 steel pipe welding

3 DN2600四管跨堤箱涵穿越北江大堤，创国内跨堤工程之最
Culvert of four DN2600 pipe crosses the Beijing River Dike, topping the country's dike-crossing projects

4 DN3600双管沉管穿越北江，创国内最大管径，单管长度最长纪录
Dual DN3600 immersed pipe crosses the Beijing River, setting the records of the largest pipe diameter and the longest single pipe in China

5 开创性的DN3600PCCP管平板车托管运输
Innovative DN3600PCCP flatbed transport

6 大型内衬钢管盾构隧道始发井，开创供水穿越障碍物的新方式
Launch shaft of shield tunnel using large lined steel pipes offers a new way for water supply pipes to penetrate obstacles

7 大型管道盾构内衬钢管穿越
Penetration of lined steel pipes in shield construction of large-diameter pipe works

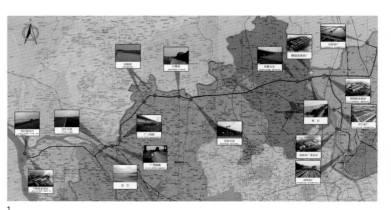

1

广州市花都区新华污水处理厂(一、二期)提标改造工程
Xinhua Sewage Treatment Plant (Phase I and II) Upgrading and Renovation Project in Huadu District, Guangzhou

新华污水处理厂是广州市花都区最重要的污水处理厂,承担了花都区50%以上生活污水处理任务;新华污水处理厂一、二期工程于2009年建成投产,总处理能力20万立方米/天;为把广州市打造成青山绿水的宜居城市,广州市政府制订了污染物减量的多项计划,本工程作为其重要的组成部分。

本提标改造工程采用活性砂滤池工艺,处理效果稳定,水头损失小;经过多轮的方案论证及实地考察,采用了活性砂滤池处理工艺并建成了目前国内最大规模活性砂滤池。

本工程运行以来,运行效果良好,出水水质稳定,污染物排放减量效果明显,得到了业内的广泛认可,接待了全国各地专家的参观考察,起到了良好的示范效应与广告效益。

As the most important sewage treatment plant in Huadu District, Guangzhou, the project undertakes over 50% of domestic sewage treatment in Huadu District. The Phase I and II project were completed and put into operation in 2009, with a total capacity of 200,000 m³/d. Guangzhou Municipal Government developed a number of plans to reduce pollutants and develop a green and livable city, while the Project is an important part of these plans.

The Project adopts active sand filter process to ensure stable treatment effect and minimum water loss. Upon several rounds of feasibility study and field investigation, active sand filter process was adopted and the largest active sand filter in China was built.

Since its operation, the Project has been running well with stable quality of treated water and remarkably less pollutant emission. It has been widely recognized in the industry and visited by experts from all over the country, playing a demonstration and publicity role.

项目地点:广州市花都区
设计时间:2012.12
建设时间:2014.4
建筑面积:设计规模20万m³/d
曾获奖项:2015年度全国优秀工程勘察设计行业奖市政公用工程三等奖
2015年度广东省优秀工程勘察设计奖一等奖

Location: Huadu District, Guangzhou
Design: 2012.12
Construction: 2014.4
GFA: 200,000 m³/d as designed
Awards: The Third Prize of National Excellent Engineering Exploration and Design (Municipal Public Works), 2015
The First Prize of Excellent Engineering Design of Guangdong Province, 2015

1

2

1 总平面图
 Site Plan

2 功能分区布置明确，打造园林式污水处理厂
 Clear functional zoning ensures a garden-style sewage treatment plant

3 污水处理与反冲洗同步进行，保障污水厂连续运行
 Simultaneous sewage treatment and backwashing ensures continuous operation of the sewage plant

4 构筑物结构与安装简洁、明了，污水处理效果良好
 The structure and installation are simple and clear, with satisfactory sewage treatment result

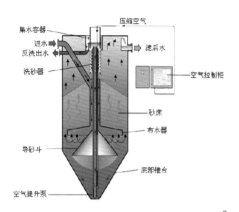

3

4

广州市江高-石井污水处理系统工程
Guangzhou Jianggao – Shijing Sewage Treatment System

广州市江高—石井污水处理系统工程是广州市迎"亚运"重点市政配套工程。该工程的建设是为了有效保护流溪河饮用水水源，改善白云区北部的生态环境。工程在满足运行稳定、处理达标等的前提下，力争建设成广州市污水处理行业的样板工程。

工程由厂外污水收集系统和污水处理厂二部分组成。工程纳污范围包括了白云区石井街及江高镇的大部分区域，纳污面积159平方公里，服务人口81.45万人，厂外污水收集系统包括提升泵站3座，截污管道152公里。厂外污水收集系统的布置充分考虑了范围内主要污染源的收集，克服了地质条件复杂多变的困难，针对性的选取合适的管道基础处理方式，取得了良好的效果。石井污水处理厂设计规模30万立方米/天，污水处理厂的设计很好地解决了厂区受征地条件限制，厂区用地地形复杂，平面布置难度大的困难。在保证使用功能的前提下解决了以下问题：① 保持山体自然景观作为厂区绿化，让构筑物布置尽量紧凑。② 厂前区与生产区利用地形高差，彻底分开为完全独立的空间，群山环绕，体现了以人为本的设计理念。③ 污泥区与预处理区集中布置，远离生活区。④ 一、二期衔接合理，一期工程相对完整，便于运转。石井污水处理厂设计采用改良型A2/O处理工艺，出水达到国家一级A处理标准。

本项目的建设，为提升广州市的污水收集及处理能力，减少初期雨水溢流污染作出了重要贡献。

Guangzhou Jianggao-Shijing Sewage Treatment System is a key municipal utility project to support the 2010 Asian Games in Guangzhou. The system aims to effectively conserve the Liuxi River drinking water source and improve the eco-environment in the north of Baiyun District. With steady operation and standard-compliant treatment, the project has set up an example in Guangzhou's sewage treatment industry.

The project is made up of two parts, i.e. the off-site sewage collection system and Shijing Sewage Treatment Plant. It serves a population of 814,500 on an area of 159km^2, covering most part of Shijing Sub-district and Jianggao Town in Baiyun District. The off-site sewage collection system has 3 pump stations and 152km of sewage intercepting pipelines. It works to the desired effect with the elaborately-devised layout to facilitate the collection of main sewage sources in the area and the proper pipeline foundation treatment to tackle the challenges brought by the diverse geological conditions. With a designed capacity of 300,000m^3/d, the plant is planned to properly tackle the site restraints, i.e the topographical complexity and the difficulty in locating all functions at one same level. While the functionality is ensured, the following issues are properly addressed in design: 1.Retaining the natural scenery as the Plant's landscaping and providing a compact building layout. 2.Separating the front area from the production area by leveraging on the topographical height difference and, with the surrounding green mountains, realizing the human-oriented concept. 3.Centralizing the sludge area and the pre-treatment area and keeping them far away from the living area. 4.Ensuring smooth connection between Phase I and Phase II, and the relative integrity of Phase I for convenient operation. The Plant adopts the modified treatment technology A2/O, and its effluent conforms to national Level 1-A standard.

The Project has contributed significantly to improving the sewage collection and treatment capacity of Guangzhou while reducing the pollution caused by initial rainwater overflow.

项目地点：广州市白云区
设计时间：2004.5
建设时间：2009.5
建筑面积：设计规模30万m^3/d
曾获奖项：2013年度全国优秀工程勘察设计行业奖 市政公用工程给排水二等奖
2013年度广东省优秀工程勘察设计奖 工程设计一等奖

Location: Baiyun District, Guangzhou
Design: 2004.5
Construction: 2009.5
GFA: designed capacity 300,000 m^3/d
Awards: The Second Prize of National Excellent Engineering Exploration and Design Award (Water Supply/Drainage of Municipal Public Works), 2013
The First Prize of Excellent Engineering Exploration and Design Award (Engineering Design) of Guangdong Province, 2013

1 总平面图
   Site plan

2 生化池采用全封闭除臭，池顶绿化与全厂绿化协调一致
   The biochemical tank is fully closed for deodorization and the tank top is landscaped to harmonize with that of the plant

3 采用周进周出圆形沉淀池
   The round settling tank features peripheral feed and peripheral overflow

4 滤布滤池与出水仪表间合建
   The cloth-media filtration tank is integrated with the effluent meter room

5 布局简约合理，外观简洁，绿化涵盖整个污水厂
   The plant features concise and effective layout, unsophisticated appearance and generous greenery

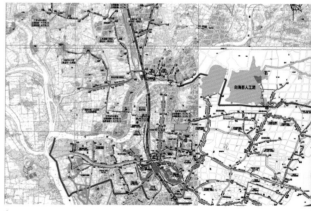

1

2

3

4

5

水处理与环保技术

GDAD水处理与环保技术研发中心（以下简称"水研发中心"）成立于2009年，由我院整合水处理领域的优秀人才组建而成，目前是省级科研平台广东省水环境与生态工程技术研究开发中心的执行管理机构。开展业务有政策研究、水处理技术研发、PPP咨询服务及创新技术设计应用等四大业务板块。

在政策制定与研究方面，GDAD积极参与省市行业主管部门的政策研究和制定，承担了省住建厅和多个地市关于污水、供水及垃圾处理等方面的法规及规划的制定，先后完成了《广东省市政公用事业特许经营管理办法》、《粤东西北地区新一轮生活污水处理设施建设实施方案》、《广东省城乡生活污水处理"十三五"规划》、《广东省城市黑臭水体整治实施方案》、《粤东西北地区新一轮生活污水和垃圾处理设施PPP模式建设操作指南》、广东省农村生活污水处理适用技术和设备指引、农村污水处理技术指引、东莞市供水行业标准化管理体系构建研究（含15个法规文件）等的编制工作。

在PPP咨询服务方面，GDAD团队致力于新时期政府和社会资本合作的政策研究，为政府和社会资本提供PPP项目全过程的咨询服务。我们是全国最早（2004年）开展公用设施特许经营招商咨询的单位之一，并一直与全国第一个特许经营权项目（来宾电厂B厂）的法律顾问单位——北京市中咨律师事务所合作，已形成一支依托我院多专业高级人才优势，由技术、投融资律师和财务等高级人才组成的PPP咨询团队，熟悉项目前期、设计、建设和运行管理全过程，注重项目需求研究和绩效考核，已在污水、垃圾等行业上创新了各种增效机制。目前开展的PPP咨询领域包括：污水处理工程、环境卫生、供水工程、黑臭水体、海绵城市、园林景观、综合管廊、特色小镇、园区开发、道路工程、公共建筑和城镇化等。

在科技创新与研发方面，经广东省科技厅2015年批准，GDAD与哈尔滨工业大学联合成立了广东省水环境与生态工程技术研究中心，形成了由工程院院士、设计大师为学术带头人的高素质研发人才队伍。工程技术研究中心与哈尔滨工业大学、同济大学、华南理工大学、广东工业大学、中科院能源研究所等科研机构合作，致力于开展水环境与生态的重大技术、关键技术、共性技术和核心技术的科技攻关和产业化研究开发，涵盖城市低影响开发、安全饮用水处理、生态城市规划、黑臭水体及河涌整治、水体污染生态控制、污泥处置、高浓度废水等方面，开发出有市场前景和竞争力的新技术、新工艺、新产品，提高企业自主创新和科技成果转化能力。

在创新技术设计应用方面，GDAD致力于创新技术的设计应用，以技术研发成果为基础，结合实践经验，完成了多个项目的创新设计，取得了良好的经济社会效益。先后完成了广州新塘水厂技术改造工程（70万立方米／天）、花都区石塘供水加压站及配套管道工程（17万立方米／天）、中山市三乡新水厂排泥水处理系统工程（10万立方米／天）、向汽车城供水管道工程（DN600~DN1400，PE管，12公里）、江门市区应急备用水源及供水设施工程（22万立方米／天，DN1400~1600管道18公里）、广州北江引水工程-花都水厂（100万立方米／天）、渗滤液处理、农村供水等。

Water Treatment and Environmental Protection

Research and Development Center of Water Treatment and Environmental Protection Technology of GDAD (RDCWTEPT) was established in 2009, staffed with a group of professionals in the water treatment field. As the executive management organization of Guangdong Provincial Research and Development Center of Water and Ecological Engineering Technology, a provincial-level scientific research platform, RDCWTEPT is currently engaged in a service portfolio of policy research, water treatment technology research and development, PPP consultancy, and innovative technology design and application.

In terms of policy-making and research, we have proactively participated in the policy research and formulation of the industrial competent authorities at the provincial and municipal levels, and formulated the laws and plans of Guangdong Department of Housing and Urban-rural Development and multiple municipal governments regarding sewage, water supply and waste treatment such as *Administrative Measures of Guangdong Province on Franchise of Municipal Utilities, Implementation Scheme for a New Round of Domestic Sewage Treatment Facility Construction in the East, West and North Areas of Guangdong Province, Thirteen Five-Year Plan of Guangdong Province for Urban and Rural Domestic Sewage Treatment, Implementation Scheme of Guangdong Province for Improvement of Urban Black and Odorous Water Bodies, Operation Guideline on a New Round of Domestic Sewage Treatment Facility Construction in PPP Mode in the East, West and North of Guangdong Province*, guidelines of Guangdong Province on applicable technologies and equipment for rural domestic water treatment, technical guidelines on rural sewage treatment, and the research on the establishment of the standardized management system for the water supply industry in Dongguan City (including 15 legal documents).

In terms of PPP consultancy, we specialize in the policy research for cooperation between the government and social capital in the new era, providing a full-process consultancy on PPP projects for the government and social capital. As one of China's earliest institutions (2004) which offer consultancy on attracting investment in utility franchise, we have been cooperating with Beijing Zhongzi Law Office, the legal counsel of China's first franchise project (Laibin Power Plant B), and, based on our multi-disciplinary talent team, developed a PPP consultation team comprising seasoned technical experts, investment and financing lawyers and accountants. This team has a good knowledge about the whole project process from the initial preparation, design, construction to the operation management, and has innovated various performance enhancement mechanisms in the sewage and waste treatment through project demand study and performance assessment. The existing PPP consultancy covers such fields as sewage treatment, environmental health, water supply, black and odorous water body, sponge city, landscaping, underground pipe gallery, characteristic town, park development, road works, public buildings and urbanization.

In terms of scientific innovation and R&D, GDAD and Harbin Institute of Technology jointly established Guangdong Water Environment and Ecological Engineering Technology Research Center(the Center) upon approval by Guangdong Provincial Department of Science and Technology in 2015, developing a high-quality R&D talent team led by the members of Chinese Academy of Engineering, and design masters. The Center works with Harbin Institute of Technology, Tongji University, South China University of Technology, Guangdong University of Technology, and Energy Research Institute, Chinese Academy of Sciences to tackle hard-nut problems and conduct industrialization R&D in major, key, common and core technologies of water environment and ecology, covering low-impact urban development, safe drinking water treatment, ecological urban planning, improvement of black and odorous water body and rivers, ecological control of water pollution, sludge disposal, and high-concentration waste water. The goal is to develop competitive new technologies and products with great market outlook and enhance the corporate ability in independent innovation and commercialization of scientific results.

In term of the design and application of innovative technology, we have completed the innovative design for multiple projects with good economic and social benefits based on technological R&D results and practical experiences. Such projects include the Technical Improvement of Guangzhou Xintang Water Plant(700,000m^3/d), Shitang Water Supply Pressure Station and Supporting Pipe Works(170,000m^3/d)in Huadu District, Sludge Water System of New Water Plant, Sanxiang Town, Zhongshan City(100,000m^3/d), Pipes of Water Supply to Motor City (DN600-DN1400, PE pipe, 12km), Emergency Backup Water Source and Water Supply Facilities in the urban area of Jiangmen(220,000m^3/d, DN1400-1600 pipe,18km), Guangzhou Beijiang River Diversion Works-Huadu Water Plant(1 million m^3/d), filtered liquid treatment and rural water supply.

花都水厂工程
Huadu Water Plant

花都水厂工程是为解决花都区乃至广州市北部供水安全性问题而实施的大型跨区调水工程，水厂取水点位于清远北江飞来峡水利枢纽附近，通过管道将原水输送至芙蓉嶂水库附近，水厂总设计规模为100万立方米／天，首期设计规模为60万立方米／天。

通过对北江取水口处水源水质的详细分析，结合国内外供水行业的发展现状和规划，本工程水厂远期制水工艺路线采用"生物预处理→混凝沉淀→超滤膜→碳吸附滤池"。其中，超滤膜技术是一种以精过滤形式去除浊度和微生物的绿色物理分离技术，属于第三代供水水处理技术，解决了第一代常规工艺（混凝–沉淀–过滤–氯消毒）及第二代深度处理工艺（第一代工艺+臭氧—颗粒活性炭）无法完成的新标准两虫（贾第鞭毛虫和隐孢子虫）等耐氯微生物问题，是饮用水处理技术的发展方向。该工艺路线既能够满足未来供水行业水质标准的进一步提高的要求，又能应对水源水质恶化等的不确定因素和风险。水厂近期制水工艺路线采用以"混凝沉淀→超滤膜过滤"为核心的处理工艺，该工艺能大大提高水厂出水的生物安全性与化学稳定性，可将出水浊度控制在0.1NTU以下，出水水质全面优于《国家生活饮用水卫生标准》的106项规定。

为应对北江水源可能出现的突然水质污染事故，本项目针对性地设计了多种药剂多点投加的药剂投加系统。本工程建成后，将成为亚洲最大的超滤膜自来水厂，对水处理行业新技术的推广应用具有巨大的推动作用和借鉴意义。

Huadu Water Plant is a large scale cross-regional water diversion project to ensure the water supply security in Huadu District and even the whole northern part of Guangzhou City. The water intake location is near the Feilaixia Water Conservancy on Beijiang River in Qingyuan City. The raw water is sent to Furongzhang Reservoir through pipeline. The designed capacity of the Plant is 1 million m³/d and Phase I, 600,000 m³/d.

Through detailed analysis of the water quality of the Beijiang water intake, as well as the latest development and trend of domestic and international water supply industry, the long-term water production process of the Plant is designed to be *biological pretreatment - coagulation sedimentation - ultrafiltration membrane - carbon adsorption filter pool*, among which, the ultrafiltration membrane technology is a green physical separation technology to remove turbidity and microbes with fine filter. As one of the third generation water treatment technologies, ultrafiltration membrane technology is able to remove chlorine-resistant microbes, especially the two parasites (Giardia and Cryptosporidium) underlined in the new standards, which cannot be removed by the first generation of conventional technology (coagulation - sedimentation - filtration - chlorine disinfection) or the second generation of deep processing technology (the first generation of technology + ozone - granular activated carbon). It represents the future direction of development of drinking water treatment technology. The designed approach is not only ready to meet the future higher requirements on water supply industry, but also able to cope with water source quality deterioration and other uncertainties and risks. The water plant takes *coagulation sedimentation - ultrafiltration membrane filtration* as core water treatment processes currently, which can greatly improve the biosafety and chemical stability of the treated water, with turbidity controlled under 0.1 NTU, better than the national standards for drinking water in all 106 items.

As countermeasures to tackle the potential emergency water pollution accidents in the Beijiang water source area, the project has designed a multiple-location chemical feeder system. Upon completion, the Plant will become Asia's largest ultrafiltration membrane water plant and play an enormously important role in promoting and demonstrating the application of new technologies in water treatment industry.

项目地点：广州市花都区
设计时间：2014
建设时间：2018
设计规模：远期：100万m³/d；首期：60万m³/d
占地面积：远期：650亩；首期：408亩
净水工艺：远期："生物预处理+混凝沉淀+超滤膜+碳吸附滤池"处理工艺；
首期："混凝沉淀+超滤膜过滤"处理工艺

Location: Huadu District, Guangzhou
Design: 2014
Construction: 2018
Capacity: Long-term: 1 million m³/d; Phase I: 600,000 m³/d
Site: 650 mu; Phase I: 408 mu
Water Treatment Processes:
　Long-term: biological pretreatment + coagulation sedimentation + ultrafiltration membrane + carbon adsorption filter pool;
　Phase I: coagulation sedimentation + ultrafiltration membrane filtration

1　总平面图
　　Site plan

2　水厂采用"生物预处理+混凝沉淀+超滤膜+碳吸附滤池"的先进处理工艺；总设计规模为100万立方米／天，首期规模为60万立方米／天
　　Advanced processes: biological pretreatment + coagulation sedimentation + ultrafiltration membrane + carbon adsorption filter pool are adopted; designed total capacity is 1 million m³/d and Phase I, 600,000 m³/d.

3　花都水厂净水工艺流程图
　　Flow Chart of water treatment process

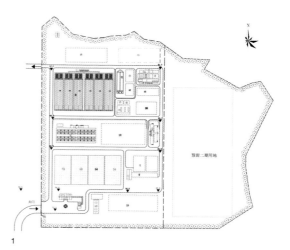

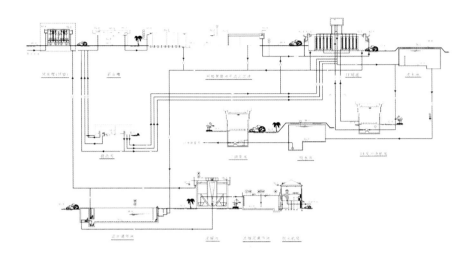

江门市区应急备用水源管道及供水设施工程
Jiangmen Urban Emergency Backup Water Supply Pipelines and Facilities

本项目建设地点位于广东省江门市，工程内容包括应急备用水源管道、那咀水库取水泵房及西江取水泵房建设。应急备用水源管道由那咀水库引DN1400管道途经杜阮西路、杜阮北三路、杜阮北二路、杜阮北一路、篁庄大道西沿线、江门大道、群星路、建设三路、北环路、滨江大道至西江水厂，管道总长为18.796公里，项目总引水规模为22万立方米／天。

为解决江门市区供水水源单一的风险，提高供水安全性，项目将那咀及那围两个水库联合调度作为西江水厂的备用水源。在极枯年份，那咀水库存水量低于应急备用所需水量时还可以由西江泵房从西江提水进行补充。在西江水源污染发生时，库区蓄水经原水管道送至西江水厂，使西江水厂成为双水源水厂，满足蓬江、江海两区应急用水需要，保障了中心城区居民生活用水，对提高城区供水保障水平和抗风险能力具有重大意义。

本工程的建设模式采用政府和社会资本合作模式(PPP模式)，特许经营期为11年，总投资约为2.75亿元，是广东省首批PPP示范项目和国家财政部第二批PPP示范项目。我院同时作为该项目的PPP咨询机构，在较短时间内协助实施部门按照省住建厅的要求顺利完成项目的总体策划和社会资本采购工作，得到政府实施部门的肯定。该项目的成功实施，为江门以PPP模式推进服务项目投资建设提供了宝贵的经验借鉴。

Located in Jiangmen, Guangdong Province, the Project includes the construction of emergency water supply pipeline, Nazui reservoir water intake and Xijiang water intake pump houses. The emergency water supply pipeline extends with DN1400 pipes from Nazui Reservoir to Xijiang Water Plant, along the roads of Du Ran Xi Lu, Du Ruan Bei San Lu, Du Ruan Bei Er Lu, Du Ruan Bei Yi Lu, Huang Zhuang Da Dao Xi, Jiangmen Da Dao, Qun Xing Lu, Jian She San Lu, North Ring Road, Bingjiang Da Dao, measuring a total length of 18.796km and a total water diversion capacity of 220,000 m^3/d.

As solution to the risky single water supply source in the urban area of Jiangmen and measure to improve water supply security, the Project takes Nazui and Nawei reservoirs as spare water sources for Xijiang Water Plant. In very dry years, when the water storage capacity in Nazui Reservoir is lower than the required amount for emergency backup, water can be taken from Xijiang Intake as supplement. When pollution occurs to Xijiang water source, the water from reservoirs will be sent to Xijiang Water Plant through raw water pipelines, to secure the Water Plant with dual water sources and meet the emergency water demands in Pengjiang and Jianghai Districts. The Project is of great importance in securing domestic water demands of downtown residents and in improving water supply security level anti-risk ability in downtown area.

The Project adopts Public-Private Partnership (PPP model) mode, with franchise period of 11 years and a total investment of about RMB 275 million. It is one of the first PPP demonstration projects in Guangdong Province and one of the second PPP demonstration projects under the Ministry of Finance. As the PPP advisory body for the Project, we has assisted the implementation authorities in completing the overall planning and social capital procurement works as required by the Provincial Department of Housing and Urban-Rural Development in a relatively short period of time, which was highly praised by the implementation authorities. The success of the Project provides valuable experiences in investment and construction under PPP model in Jiangmen.

项目地点： 江门市
设计时间： 2015.10
建设时间： 2016
设计规模： 22万m^3/d
管道总长： 18.8km
管径及管材：管径: DN1400; 管材: 球墨铸铁管

Location: Jiangmen
Design: 2015.10
Construction: 2016
Capacity: 220,000 m^3/d
Total Length of Pipelines: 18.8 km
Pipe Diameter and Material: Diameter: DN1400; Material: Ductile Iron Pipe

1 那咀水库取水泵房总平面图
  Site Plan of Nazui Reservoir Water Intake Pump House
2 应急备用水源管道总平面图
  Site Plan of Emergency Backup Water Source Pipeline
3 天沙河管线架空段施工现场
  Site of Tiansha River Pipeline Overhead Section
4 杜阮北一路段DN1400球墨铸铁管吊装施工现场
  Site of Lifting DN1400 Ductile Iron Pipe in Du Ran Bei Yi Lu Section
5 那咀水库取水泵房栈桥施工现场
  Site of the Pier of Nazui Reservoir Water Intake Pump House
6 杜阮北一路段DN1400球墨铸铁管吊装施工现场
  Site of Lifting DN1400 Ductile Iron Pipe in Du Ran Bei Yi Lu Section
7 天沙河管道架空段施工图
  Site of Tiansha River Pipeline Overhead Section

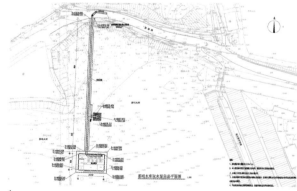

1

2

3 4 5

6

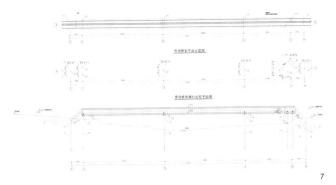

7

郁南县整县生活污水处理捆绑PPP项目
PPP Bundled Project for Sewage Treatment Plants of Yunan County

郁南县在2015年被住建部选入全国100个农村污水治理示范县，成为粤东西北地区唯一的全国农村生活污水治理示范县，同时也是我省首个采用整县污水捆绑打包PPP创新模式的县城。

郁南县整县生活污水处理捆绑PPP项目将整县15个乡镇及农村地区生活污水处理设施建设内容合理整合，其中县城区和两个中心镇（连滩镇、南江口镇）通过续建污水收集管网，完善污水收集系统，提高污水处理设施效能；12个乡镇通过新建镇一级污水处理设施及配套管网，满足当前镇区污水处理的需求，填补镇级污水处理设施缺失的短板，实现乡镇一级污水处理设施全覆盖；通过连片整治的方式完成全县共903个农村的生活污水处理设施建设，改善农村污水处理现状，提升卫生条件；另外，该项目还在15个乡镇分别挑选出一个中心村作为试点实施人居环境综合提升的建设内容，从基础设施建设到产业开发全方位提升农村人居环境质量。该项目镇区和农村共新建污水处理规模达1.95万吨/日，配套污水收集管网长度达73.7公里，总投资约5.02亿。

郁南项目采用竞争性磋商方式进行社会资本采购，该项目发布上网后即吸引了国内多家水务大型企业的关注，最终共有7家水务行业的龙头企业参与竞争，从发布资格预审公告到完成社会资本采购，共经历两轮磋商环节耗时4个月，最终年服务费成交价较政府预期下浮30%，根本上取得了降低项目全生命周期成本、减少回报的最佳效果。实践证明，"厂网捆绑，进出水质考核"的项目条件被水务资本市场接受，郁南项目创立了村镇污水产业PPP新模式，是全国第一个整县污水捆绑打包、采用环境污染减排付费的PPP项目，彻底解决我国污水收集浓度不高的关键问题。

项目地点：云浮市郁南县
设计时间：2016.1
建设时间：2016.7

Location: Yunan County, Yunfu City
Design: 2016.1
Construction: 2016.7

Yunan County was selected by the Ministry of Housing and Urban-Rural Development in 2015 as one of the 100 rural sewage treatment demonstration counties. It is the only one of its kind in the rural areas in eastern, western and northern Guangdong and the first county to adopt the innovative PPP model of the whole county sewage treatment in Guangdong.

The PPP project consolidates the construction demands in a systematic way for domestic sewage treatment facilities in 15 townships and rural areas in the whole county. In the system, the sewage collection system in the county downtown area and two center towns (Liantan Town and Nanjiangkou Town) are improved by extending sewage collection pipe network to increase the efficiency of sewage treatment facilities, while the demands for sewage treatment of the other 12 townships are met by building new sewage treatment facilities and pipe networks at township level, filling the blank and achieving full coverage of sewage treatment facility at township level. Contiguous improvement methods are implemented in construction of domestic sewage treatment facilities serving 903 villages under the administration of the county to improve the status quo of rural sewage treatment and health conditions. In addition, the Project has also selected one center village in each of 15 townships as the pilot for the comprehensive improvement plan of living environment, which helps improve infrastructure, industrial development and living environment quality in rural area in a comprehensive way. The Project has newly built a capacity of sewage treatment of 19,500t/d and total sewage collection pipeline of 73.7km with a total investment of RMB502 million.

The Project called for social capital procurement through competitive negotiation, attracting extensive attention from a number of large domestic water treatment enterprises since the announcement on internet. Finally seven of them participated in the competition and it took two rounds of negotiation in four months from the release of prequalification notice to the completion of social capital procurement. The finally settled annual service fee is 30% lower than the government expectation, achieving the best effect of reducing the life cycle cost and return of the Project. It is proven that the conditions, such as bundled construction of water plant and pipe network and water quality assessment on untreated and treated water, set by the Project were well accepted by the water treatment capital market. In this sense, the Project has set a new PPP model for rural sewage industry. Besides, as the country's first PPP project that bundles the whole county's sewage treatment into a package and takes effluent reduction as payment means, the project has offered a solution to the low concentration of sewage collection, a key problem facing the country.

分批次规划污水处理设施子项目实施顺序，有序推进整县污水处理设施建设

Phased planning of sub-projects ensures orderly progress of construction of sewage treatment facilities in the whole county

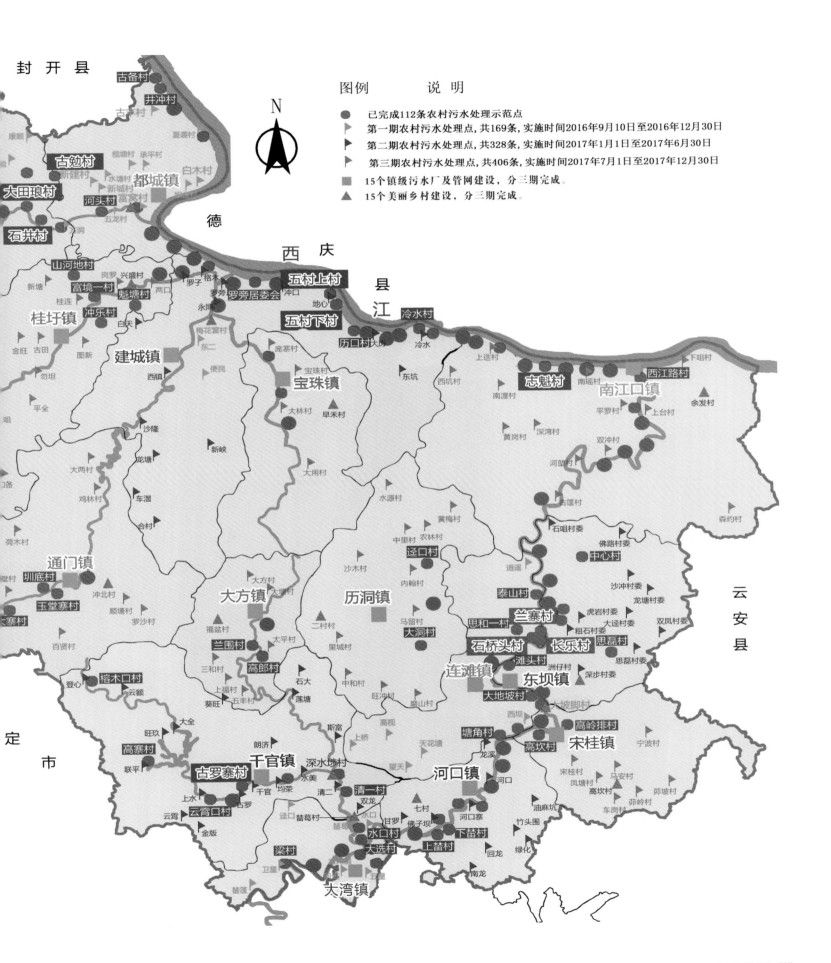

环境保护

环保专项业务范围涵盖生活垃圾、工业固废、污泥、粪便、城市污水、工业废水处理等市政、环保工程领域等。本环保章节中，主要列举了我院近年来已经完成的几项环保工程作品，以供参考交流。

台山市台城下豆坑生活垃圾卫生填埋场采用无害化卫生填埋处理方式和高维填埋的设计方法，提高库容量和使用年限，防渗系统采用HDPE膜+GCL膨润土垫的防渗系统保护地下水免受污染。项目总占地697.967亩，总库容约850万立方米，分三期工程建设，使用年限约50年，设计规模为日处理生活垃圾400吨。工程第一期总投资约10925万元，规划使用地180亩，分二区实施。该项目相继获得"广东省一级无害化填埋场"、"广东省生活垃圾处理示范项目"、"广东省优秀工程勘察设计三等奖"等多个奖项。

惠州市惠阳区榄子垅环境园生活垃圾综合处理场BOT项目总控制用地约1138.91亩，投资5.1亿元。项目建设成集生活废弃物处理利用、技术科研、宣传教育于一体，突出生态、环保理念的"环境园"，在园内配套建设垃圾卫生填埋场、垃圾焚烧发电厂、金属分选及环保制砖厂、粪便处理、餐厨垃圾、建筑垃圾处理、污泥处理、垃圾分拣中心、沼气发电等一系列垃圾处理设施，充分利用各项目之间的协同效应，实现各项目资源的二次开发和循环利用，达到节能减排的目的，提高综合效益，具有示范效应。

佛山高明苗村白石坳垃圾卫生填埋场渗滤液处理厂二期工程是佛山市环卫事业重点项目，本项目渗滤液处理采用"外置式膜生物反应器（MBR）+两级纳滤（NF）"作为渗滤液处理主体工艺，MBR 中的生化系统采用两级AO 系统，膜深度处理系统采用两级纳滤工艺，在保障了系统出水达标排放的同时具有节约运行成本、产水率高等优点。浓缩液处理采用"三级纳滤（NF）+斜板蒸发+水泥固化"的处理工艺，浓缩液进行深度处理后产水达标排放，解决了传统浓缩液回灌带来的一系列影响填埋场运营的问题。

广州市餐厨废弃物循环处理试点项目是广州市首个实施的餐厨废弃物资源化项目，也是广东省内首个采用好氧生化工艺将餐厨废弃物转化为微生物菌剂作农业肥的项目，具有示范意义。项目采用微生物发酵技术工艺，充分利用餐厨废弃物中的营养成分，通过微生物好氧发酵工艺生产农用微生物菌剂。该技术及时、彻底解决了餐厨废弃物无害化处理问题，减少人畜交叉感染和环境污染，以工业化方式生产生物腐殖酸肥料，不仅解决了城市餐厨废弃物处理难题，而且从源头杜绝食品安全问题。

丰顺县县城五斗种城市生活垃圾卫生填埋场改扩建工程采用高维填埋的设计方法，提高库容量和使用年限。项目设计容积约为97.38万立方米，服务年限约为14.7年。本项目防渗系统采用1.5毫米HDPE膜和4800克／平方米膨润土防水垫层的复合防渗衬层结构保护地下水免受污染。填埋场设计建设有先进的渗沥液处理系统、气体燃烧排放系统、地下水污染预防系统、防尘防臭味扩散措施及防噪声措施等。

Environmental Protection Projects

Our environmental protection business covers the treatment of domestic waste, industrial solid waste, sludge, excrement, urban sewage and industrial sewage in municipal and environmental protection fields. Below are several environmental projects completed by GDAD in recently years for the readers' reference.

Taicheng Xiadoukeng Domestic Waste Sanitary Landfill features harmless sanitary landfill and high-order landfill for larger storage capacity and longer service life and anti-seepage system made of HDPE membrane + GCL (geosynthetic clay liner) to protect groundwater from pollution. Occupying a land of 697.967 mu, the three-phase Project is designed with the total storage capacity of 8.5 million m^3, the service life of 50 years and daily waste treatment throughput of 400 tons. The Phase I project on 180-mu of land costs around RMB 109.25 million and is implemented in two zones. The Project has won the awards including First Grade Harmless Landfill of Guangdong Province, Demonstration Project of Domestic Waste Treatment of Guangdong Province and Third Prize of Excellent Engineering Exploration and Design Award of Guangdong Province.

Lanzilong Environmental Park Domestic Waste Comprehensive Treatment Plant, a BOT Project in Huiyang District, Huizhou, is the first Environmental Park proposed and implemented within the administrative region of county-level city (district) in China. The overall plan and phased development plan of the whole park are formulated at the project establishment stage, offering a feasible operation mode and basis of project establishment for other similar projects. The Park integrates the treatment and utilization of domestic wastes, technical research and publicity and education with highlighted eco- and environment-friendly concept. With a total land area of 1,138.91mu, the Project is developed through overall planning and land acquisition and phased development. The Park is equipped with waste sanitary landfill, waste incineration power plant, metal separation and environment-friendly brick factory and treatment facilities for excrement, cooking waste, building waste, sludge, waste sorting center and biogas power generation. The synergy between different facilities is fully leveraged to realize the secondary development and recycling of different resources for energy conservation and emission reduction, which enhances the all-around benefits and exemplary efforts.

Leachate treatment plant (Phase II) under Bai Shi Ao Waste Sanitary Landfill in Miao Village, Gaoming District, Foshan City is one of the key projects of environmental hygiene sector of Foshan. The external MBR + two-stage NF is employed as the main technology of leachate treatment. MBR is provided with two-stage AO system as the biochemical system and carbon dosing device for improving the denitrifying capacity and adaptation of biochemical system to leachate at different time; the advanced treatment of membrane adopts two-stage NF to ensure up-to-standard water discharge with lower operation cost and high water productivity; the concentrates are treated with three-stage NF+ sloping evaporation + cement solidification. After the advanced treatment, the up-to-standard water is discharged and a small amount of vaporized concentrates is subject to the safe and sanitary landfill after the cement solidification. The proper treatment of the concentrates solve the problem of the landfill operation brought by the conventional leachate recirculation.

Guangzhou Cooking Waste Recycling Pilot Program is the first project in Guangzhou to resource the cooking waste and the first project in Guangdong to transform the cooking waste into the microbial agent as agricultural fertilizer by following the aerobic biochemical process, which is of exemplary significance. With the microbial aerobic fermentation technology, the nutrient content in the cooking waste is fully transformed into the agricultural microbial agent. Such technology addresses the harmless treatment of cooking waste in a timely and thorough way to avoid the cross infection between human and livestock and environmental pollution. The industrial production of the biological humic acid fertilizer offers a favorable solution for the treatment of urban cooking waste and ensure the food safety at source; in addition, the microbial flora from recycling can be widely used in the modern urban agriculture to recycle the resource in a real sense. The Project features matured technology, high-level automation, small site, independent deodorization and dust collection and marketable product.

For the renovation and expansion of Wudouzhong Urban Domestic Waste Sanitary Landfill in Fengshun County, the high-order landfill is employed for larger storage capacity and longer service life. The Project is designed with the capacity of 973,800m^3 and life service of 14.7 years. As the priority and challenge of the waste sanitary landfill, the anti-seepage system is designed with the composite anti-seepage liner structure made of 1.5mm HDPE film + 4,800g/m^2 bentonite waterproof cushion to prevent the pollution of underground water, which meets the requirement of code and lowers the cost. The landfill is provided with up-to-date leachate treatment system, gas combustion and emission system, underground water pollution prevention, dust prevention and deodorization measures and anti-noise measures. The MBR (two-stage A/O+UF) + NF+RO for the leachate treatment better responds to the actual demand of Fengshun area, which is suitable to the Project following reliable operation in the Pearl River Delta.

台山市台城下豆坑生活垃圾卫生填埋场一期工程
Xia Dou Keng Domestic Waste Landfill (Phase I), Taicheng, Taishan

台山市台城下豆坑生活垃圾卫生填埋场一期工程是台山市重点民生工程项目，由台山市政府筹资建设。

本项目生活垃圾填埋场设计采用高维填埋的设计方法，提高库容量和使用年限。项目一期工程设计总容积约为292万立方米，服务年限约为22年，至2020年日均处理生活垃圾量为400吨/日，2020～2030年日均处理生活垃圾600吨。项目总占地465333平方米（合698亩），填埋一期用地120000平方米（合180亩），折合300平方米／吨，处理成本折合 34元/吨。防渗系统采用HDPE膜+GCL膨润土垫的防渗系统保护地下水免受污染。渗滤液采用"MVC蒸发+DI离子交换"工艺,该工程设计的特点：1. 该工艺具有较强的适应性和操作上的灵活性，可以适应不同时期的处理需要，处理后的渗滤液可以达到排放标准。2. 该工艺有机负荷高，抗冲击负荷能力强，进水水质对其影响较小，能耗较小。3. 渗滤液处理达标准后排放，不存在渗滤液原液外流污染地表水或地下水的问题，具有很好的环境效益。

项目自2012年投产至今，日均生活垃圾处理量约为380吨／天，最大日生活垃圾处理量达到500吨／天，5～9月份渗滤液日均产生量150立方米／天，其余月份渗滤液日均产生量60立方米／天，通过有效容积为10000立方米渗滤液调节池贮存，渗滤液日均处理能力100立方米／天，可满足工程需求。

Xia Dou Keng Domestic Waste Landfill (Phase I) Project in Taicheng of Taishan City is a government-funded key project for people's well-being.

The project employs high-order landfill technology to improve the storage capacity and service life. The Phase I project is designed with a storage capacity of 2,920,000m^3, a service life of about 22 years, and an average domestic waste treatment capacity of 400t/d by 2020 and 600t/d during 2020-2030. The project occupies a site area of 465,333m^2 (roughly 698 mu), and land area for Phase I landfill is 120,000m^2 (roughly 180 mu), being equivalent to 300m^2/t with a treatment cost of RMB34/t. Seepage control system consists of HDPE membrane +GCL bentonite layer to protect the groundwater against contamination. Leachate is treated by the technology of MVC evaporation +DI Ion Exchange. Such technology boasts the following properties: 1. The technology offers better adaptation and flexibility in operation to meet various treatment demands in different phases, and the treated leachate is up to discharge standards. 2. The technology has higher organic loading, stronger resistance to impact load, less impact from inflow water quality, and lower energy consumption. 3. Treated leachate won't be discharged until reaching standards, avoiding the problem of raw leachate outflow that pollutes groundwater or surface water and generating environmental benefits.

Since its operation in 2012, the project has realized an average and maximum daily domestic waste treatment capacities of about 380t/d and 500t/d respectively. With an average daily leachate production of 150m^3/d from May to September and 60m^3/d during the other months, the average daily leachate treatment capacity is 100m^3/d and meets the project requirements thanks to the storage of a regulating tank in the effective capacity of 10,000m^3.

项目地点：台山市台城下豆坑
设计时间：2011
建设时间：2012
占地面积：占地180亩
处理规模：处理生活垃圾400～600吨/天
工程投资：1.09亿元
曾获奖项：2013年广东省一级无害化填埋场(广东省住房和城建设厅)
2015年度广东省优秀工程勘察设计三等奖
2015年广东省生活垃圾处理示范项目（广东省环境卫生协会）

Location: Xia Dou Keng, Taicheng, Taishan City
Design: 2011
Construction: 2012
Site: 180mu
Capacity: Domestic waste treatment 400-600t/d
Investment: RMB 109 million
Awards: Grade I Non-hazardous Landfill of Guangdong Province, 2013 (by the Department of Housing and Urban-rural Development of Guangdong province)
The Third Prize of Excellent Survey and Design of Guangdong Province, 2015
Exemplary Project for Domestic Waste Treatment of Guangdong Province, 2015 (by Environmental Health Association of Guangdong Province)

1

1 入口及管理区
  Entrance and administrative area

2 项目总平面布局
  Site plan

3 填埋作业区
  Landfill area

4 渗滤液调节池及渗滤液处理站
  Leachate regulating tank and leachate treatment plant

惠州市惠阳区榄子垅环境园生活垃圾综合处理场BOT项目
Lan Zi Long Environmental Park Domestic Waste Treatment Plant (BOT), Huizhou

惠州市惠阳区榄子垅环境园生活垃圾综合处理场BOT项目是惠州市重点项目。

惠州市惠阳区政府引入垃圾处理园区的概念，建设成集生活废弃物处理利用、技术科研、宣传教育于一体，突出生态、环保理念的"环境园"，项目总控制用地约1138.91亩，处理总规模为1200吨/日，投资约5.1亿元。采取总体规划、统一征地、分期建设的方式进行开发，在园内配套建设垃圾卫生填埋场、垃圾焚烧发电厂、金属分选及环保制砖厂、粪便处理、餐厨垃圾、建筑垃圾处理、污泥处理、垃圾分拣中心、沼气发电等一系列垃圾处理设施。项目充分利用各子项之间的协同效应，实现各子项资源的二次开发和循环利用，达到节能减排的目的，提高综合效益。同时在园内还将根据现有的地形地貌，设置环保教育基地、休闲山顶公园、亲民荔枝林、生态湿地公园、生态农场等一系列生态环保设施，提升整个园区环境质量。

本项目在国内县级市（区）的行政范围内首次提出"环境园"的理念，并付诸实施，在项目立项阶段即完成整个园区的总体规划和分期建设计划，为同类项目的实施提供了可行的操作模式和立项依据。此模式现已成为广东省乃至全国垃圾等固废项目建设的示范项目，如广州、中山、江门、东莞、佛山等地已陆续开展"环境园"、"循环经济园"、"静脉产业园"以及"固废环保科技园"等项目，把生活垃圾、餐厨垃圾、污泥、粪便、动物尸骸、建筑垃圾等固体废物集中到一个园区综合处理，达到资源共享、循环利用、节能减排的目的。

The project is developed as a key project of Huizhou City.

The government of Huiyang District, Huizhou introduced the concept of waste disposal park and developed an ecological and environmental protection-oriented Environmental Park that integrates domestic waste disposal and recovery, technology and research, publicity and education. The project is planned with a site area of about 1,138.91 mu, a total treatment capacity of 1,200t/d and a total investment of about RMB510 million. The project development is based on overall planning/land acquisition and phased construction. The park is equipped with a series of waste treatment facilities, including waste landfill, waste incineration power plant, metal sorting and environmental brick factory, feces disposal, kitchen waste, construction waste disposal, sludge disposal, waste sorting center, biogas power generation plant etc. By fully tapping the synergy of all its facilities, the project has realized secondary development and recycling of all its resources for energy conservation and emission reduction, thus improving the comprehensive benefits. Meanwhile, based on existing land conditions, a series of ecological and environmental protection facilities including environmental protection education base, mountain top leisure park, lychee park, eco-wetland park, and eco-farm etc. will be provided in the park to enhance the overall environment quality of the park.

It was in this project that the concept of Environmental Park was proposed and implemented at the county-level cities (districts) in China. The master plan and phased development plan of the whole park were already completed at the project establishment approval phase, offering workable project mode and basis of project establishment approval for other similar projects. The project now has become a role model for solid waste project development in Guangdong Province and even China. Such cities as Guangzhou, Zhongshan, Jiangmen, Dongguan and Foshan etc. have developed Environmental Park, Recycling Economy Park, Venous Industrial Park and Solid Waste Environmental Science Park etc.. These facilities centralize the solid wastes such as domestic rubbish, kitchen waste, sludge, feces, animal remains, construction waste etc., into one park for comprehensive treatment and eventually realize the resource sharing, recycling and recovery, energy conservation and emission reduction.

项目地点：惠州市惠阳区榄子垅
咨询时间：2011-2013
建设时间：2014
占地面积：1138.91亩
处理规模：处理生活垃圾1200吨/日
工程投资：5.1亿元
曾获奖项：2012-2013年度广东省优秀工程咨询成果一等奖
2014年度全国优秀工程咨询成果三等奖

Location: Lan Zi Long, Huiyang District, Huizhou
Consultation: 2011-2013
Construction: 2014
Site: 1,138.91mu
Capacity: Domestic waste treatment 1200t/d
Investment: RMB510 million
Award: The Third Prize of National Excellent Engineering Consultancy Deliverable Award, 2014

1 低碳馆、政府办透视
  Perspective: low-carbon hall and government office
2 办公楼透视
  Perspective: office building
3 园区透视
  Perspective: park
4 景观透视
  Perspective: landscape
5 惠阳环境园整体效果图
  Rendering: Huiyang Environmental Park

佛山高明苗村白石坳垃圾卫生填埋场渗滤液处理厂二期工程
Leachate Treatment Plant (Phase II), Bai Shi Ao Waste Sanitary Landfill, Foshan

佛山高明苗村白石坳生活垃圾卫生填埋场是2003年广东省重点工程中环保项目之一，也是当年佛山市的重点工程。佛山高明苗村垃圾卫生填埋场渗滤液处理厂二期工程是2016年佛山市环卫重点项目。

二期工程位于一期工程的南面坡地，与一期工程相隔一条7米宽的场内道路。现状坡地标高为58～81米，占地面积约10934平方米。本项目工程总投资概算为8550.68万元，项目日均进水总量640立方米／天（包括425立方米／天的渗滤液原水和215立方米／天一期工程纳滤浓缩液），最终达标清水排放量为640立方米／天。

本项目渗滤液处理采用"外置式膜生物反应器（MBR）+两级纳滤（NF）"作为渗滤液处理主体工艺，MBR中的生化系统采用两级AO系统，并设置碳源投加装置以提高MBR系统反硝化能力、强化生化系统对不同时期渗滤液的适应性；膜深度处理系统采用两级纳滤工艺，在保障了系统出水达标排放的同时具有节约运行成本、产水率高等优点。浓缩液处理采用"三级纳滤（NF）+斜板蒸发+水泥固化"的处理工艺，浓缩液进行深度处理后产水达标排放，少量的蒸发浓缩液经水泥固化后进行安全卫生填埋。浓缩液的妥善处理解决了传统浓缩液回灌带来的一系列影响填埋场运营的问题。

Bai Shi Ao Waste Sanitary Landfill in Miao Village, Gaoming, Foshan is one of the key environmental protection projects of Guangdong Province in 2003 and a key project of Foshan Municipality in the same year. Its Leachate Treatment Plant (Phase II) is a key environmental sanitation project of Foshan Municipality in 2016.

The project of Phase II, located on south hillside of Phase I, is separated from Phase I by a 7m wide internal road on site. Existing level of the hillside ranges from 58 to 81m on a site area of about 10,934m^2. With an estimated total cost of about RMB85,506,800 Yuan, the project is designed with a daily influent of 640m^3/d (including 425m^3/d raw leachate and 215m^3/d nano-filtration concentrate from Phase I) and an effluent of 640m^3/d.

The system of external MBR + two-stage NF is employed as the main technology for leachate treatment. MBR is provided with two-stage AO system, as well as additional carbon dosing device to improve the MBR's denitrifying capacity and adaptation of biochemical system to leachate at different time. The advanced treatment of membrane adopts two-stage NF to ensure up-to-standard effluent at lower operation cost and high water productivity. The concentrates are treated with three-stage NF+ sloping evaporation + cement solidification. After the advanced treatment, the effluent is discharged and a small amount of vaporized concentrates is subject to the safe and sanitary landfill after the cement solidification. The proper treatment of the concentrates solves the problem on the landfill operation brought by the conventional leachate recirculation.

项目地点：佛山高明苗村白石坳生活垃圾填埋场内
设计时间：2015
建设时间：2016
占地面积：10934m^2
处理规模：渗沥液原液425立方米／天，一期纳滤浓缩液215立方米／天
工程投资：8550万元

Location: Bai Shi Ao Waste Sanitary Landfill, Miao Village, Gaoming, Foshan
Design: 2015
Construction: 2016
Site: 10,934m^2
Capacity: Raw leachate 425m^3/d, and NF concentrate (Phase I) 215m^3/d
Investment: RMB85.5 million

1 渗滤液处理二期总平面布局
  Leachate Treatment (Phase II) Site Plan

2 渗滤液膜处理车间
  Leachate membrane treatment workshop

3 生化池池顶
  Top of biochemical pond

景观设计

景观是指土地及土地上的空间和物体所构成的综合体。它是复杂的自然过程和人类活动在大地上的烙印。它包括的内容很广，既有城市景观空间的处理；城市标志场地的设计；城市市政广场、公共绿地、道路、防护绿地的设计；城市与江、河、湖、海、山川等自然环境临界地带的设计；也包括各类人工和自然景观及小品的设计等，它涉及城市的功能、文化、技术、视觉等多方面的问题，是一个城市给予人们整体、持久印象的主要因素。本景观章节中，主要列举了我院近年来已经完成的几项景观设计作品，以供参考交流。

广州铁路新客站地区公共绿化和广场景观工程是全国四大铁路枢纽之一——广州新客站的重要配套市政项目，是一个具有多样景观空间类型的大型综合设计项目。设计师充分理解广州新火车站地区的特性环境，采用合理的设计手法打造属于该地区的城市景观空间，项目的核心景观区为新客站主体建筑前的中轴广场，设计上采用简洁、灵动、流畅的景观营造手法，创造性的采用双层式立体步行平台，合理地解决了火车站、汽车站、地铁等多种客流的交叉问题，完善了城市主干道的人行天桥系统，也创造了立体多变的景观空间效果，为广州南大门的景观环境锦上添花，给来访广州的世界各地朋友留下了深刻的"第一眼印象"。

玲玎海岸园林景观工程位于珠海东澳岛，设计者充分尊重海岛依山傍水的自然景观特色，保持并合理利用现状山地、沙滩、红石滩、相思林带等景观资源，形成地块特色明显的景观分区，打造出一个最具亚热带南国岛屿风光特色的旅游度假胜地景观。

作为珠海市第一个海岛整体开发项目，设计者与建设单位在可供借鉴的相关案例较少前提下克服技术、地质条件以及交通不便等各种因素，通力合作，共同努力，为"百岛之市"的珠海市倾情打造了一个推广海岛游的示范岛屿，也为建设单位进一步推进其他海岛旅游资源的开发，提供了宝贵的第一手数据资料。

沿着珠江支流白沙河西岸规划的金沙洲p线滨江绿地是金沙洲居住新城的唯一滨江绿地，设计以居民的步行尺度为依据布置着各种娱乐休闲设施，项目建成之后，在各市级滨江绿地中具有自己鲜明的特色，成为了金沙洲居住新城的一个地标性景观，并成为市民最喜欢去的休闲活动场所之一。

景观设计与规划、建筑设计一样是人类构筑生存环境的学科之一，而相对于规划、建筑，景观设计更关注营造适合人类生活的室外空间，它具有丰富的尺度空间，可以大到区域化的规划布局，中到城市尺度的绿地、公园、街道公共空间，也可以小到人的尺度及体验。在当今城市快速发展中生态环境日益恶化的情况下，景观设计将为实现城市青山绿水，蓝天白云，实现党十八大提出的建设"美丽中国"目标发挥着无法替代的作用。

Landscape Design

Landscape refers to a composite of land as well as the space and objects on the land. It is the mark left on the Earth by the comprehensive natural and human activities. It covers a wide range of contents from the processing of urban landscape spaces to the designs of urban landmarks, urban municipal squares, public green space, roads, green buffer, the interface where the city borders on the natural environment such as river, lake, sea and mountain, and all kinds of artificial and natural landscape and featured landscape. Landscape involves a variety of issues such as a city's functions, culture, technology and visual effects, i.e. the main factors that a city creates an overall and lasting impression. In this chapter, several landscape design works completed by GDAD in recent years are provided for reference and exchange of ideas.

New Guangzhou Railway Station boasts one of the four major national railway hubs in China. The public greening and the square landscape project there is an essential supporting municipal project for the Station. It is a large integrated design projects consisting of a variety of landscape space types, including not only the landmark central square of the Station, but also various neighborhood parks, regional road landscape, and riverside landscape. The project covers a total area of 11.4 km^2. Based on thorough understanding of the station development and planning and the surrounding environment, we employ corresponding design approaches to create the urban landscape space specific to this area. The core landscape area of the project is the central square in front of the main station building. Using simple, flexible and smooth landscaping techniques, we creatively build a double-level three-dimensional pedestrian platform which properly addresses the problem of crossing passenger flows commonly experienced in the train stations, bus stations and metro stations. The platform also becomes a part of the pedestrian bridge system over the main roads, creating a varied three-dimensional landscape space, enhancing the landscaping of the Southern Gateway of Guangzhou and leaving a favorable first impression on visitors from all around the world.

Lingding Costal Landscape Project is located in Dong'ao Island, Zhuhai. With full respect for the Island's natural landscape by the mountain and the sea, we maintain and make reasonable use of the existing landscape, such as the mountain, beach, red pebble beach and Acacia forest belt, to realize distinctive landscape zoning on the project site. While preserving and restoring the native vegetation, we select and plant fine tree species suitable for the Island and site features in the master planning, creating the most representative resort landscape of a subtropical island in south China.

As this was the first island development project in Zhuhai, there were few precedents for the client or us to learn from. Under this circumstance, we worked closely with the client to overcome the difficulties in technologies, geological conditions and inconvenient traffic. The project not only sets up an example for the promotion of the island tourism in Zhuhai, a city known as City of Hundreds of Islands, but also provides valuable first-hand data for the tourism development of other islands.

Jin Sha Zhou Line P riverside green space along the west bank of the Baisha River, the tributary of the Pearl River, is the only waterfront green space of the Jin Sha Zhou Residential New Town. Various recreation and leisure facilities are provided based on the pedestrian scale of the residents. Upon completion, it has presented distinct features as a landmark view of Jin Sha Zhou Residential New Town, and has become one of the most popular recreation and leisure venues frequented by the residents.

Landscape design, same as planning and architectural design, is one of the disciplines of building the living environment for human. Compared to them, landscape design focuses more on creating livable outdoor spaces. The space scale of landscape design varies vastly, ranging from regional planning and layout, city-scale green space, parks and street public space, to the human-scale experiences. Nowadays when the ecological environment is threatened by the drastic urbanization, landscape design will surely play an irreplaceable role in creating a clean and green urban environment and attaining the goal of building a Beautiful China proposed in the 18th CPC National Congress.

广州铁路新客站地区公共绿化和广场景观工程
Greening and Square of New Guangzhou Railway Station, Guangzhou

广州铁路新客站是全国四大铁路枢纽之一，是服务珠三角、面向华南地区的区域交通平台，是衔接珠三角城际轨道交通的客运枢纽，是广州市最为重要的窗口和门户之一。广州铁路新客站的建成满足了适应珠江三角洲区域和广州市大规模客流交通发展的需要，强化了广州市区域中心城市的地位。

本项目作为广州新客站的重要配套市政项目，是一个具有多样景观空间类型的大型综合设计项目，既有新客站配套的标志性中轴广场景观设计，也有各种街区公园、区域内道路景观、河涌沿岸滨水景观的设计。总体区域面积达11.4平方公里，项目将对整体景观环境的设计品位及未来建成后的城区环境面貌产生重要而深远的影响，其中中轴广场区除了合理解决交通功能外更肩负着对外展示广州形象的功能。

广州铁路新客站中轴广场区，是项目的核心景观区。规划构思上采用"时代性"语言、"铁路文化"要素、广州地理人文要素，塑造具有地方及铁路特色的景观形象。总体构图上采用简洁、灵动、流畅的景观营造手法，设计了灵活多变的二层步行平台，并以二层步行平台为主体，与中轴广场地面层形成连续的双标高步行系统。不同标高的步行空间用踏步、坡道、台地、楼梯等方式灵活联系，合理解决火车站、汽车站、地铁等多种客流的交叉问题，同时二层平台一系列展示大地特征的几何形体绿地及水景空间的营造，也形成了立体多变的景观空间效果。

The new Guangzhou Railway Station is one of the four major national railway hubs in China. It boasts a regional transport platform serving the Pearl River Delta (PRD) and the south China, and a passenger transportation hub joining intercity rails in the PRD. It is also one of the most important windows and gateways of Guangzhou. The Station, since its completion, meets the demands of PRD and Guangzhou for mass passenger transit, thus strengthens the status of Guangzhou as a regional central city.

As an important supporting municipal project for the Station, this Project involves various landscape spaces, including the iconic Central Square, neighborhood parks, regional road landscape and riverside landscape. Totaling an area of 11.4km^2, the Project will have far-reaching impact on the design quality of the overall landscape and the future urban environment. In particular, the Central Square area is a showcase of the city image in addition to its transportation function.

As the core of the project, the Central Square area is devised with the planning vocabulary that reflects our time, the railway culture and local geography and culture in Guangzhou, striving to create a landscape full of local and railway features. For the overall composition, concise and flowing landscaping approach is employed. A flexible F2 pedestrian platform is provided, and together with the ground level of the Central Square, forms a continuous two-level pedestrian system. Pedestrian spaces at different levels are connected via steps, ramps, platforms and staircases, a perfect solution to the problem of crossing passenger circulation commonly experienced in train stations, bus stations and metro stations. Meanwhile, the geometric lawns and waterscape on F2 platform symbolize the terrain features and create varied and multi-level landscape spaces.

项目地点：广州番禺区
设计时间：2009.2
建设时间：2012.8
建设规模：11.40km^2，核心区规模约26万m^2
曾获奖项：2013年度全国优秀工程勘察设计行业奖园林景观二等奖
2013年度广东省优秀工程勘察设计奖二等奖
2011年度获广东省优秀城乡规划设计城市规划类表扬奖

Location: Panyu District, Guangzhou
Design: 2009.2
Construction: 2012.8
Size: 11.40 km^2; core area of about 260,000 m^2
Awards: The Second Prize of Garden & Landscape under National Excellent Engineering Exploration and Design Award, 2013
The Second Prize of Excellent Engineering Design Award of Guangdong Province, 2013
Testimonial Prize (Urban Planning) of Outstanding Urban and Rural Planning and Design Award of Guangdong Province, 2011

1 总平面图
  Site plan
2 鸟瞰图
  Bird's eye view
3 故乡水叠石景观
  Waterscape with stacked rockery
4 清朗、祥和、自然的河涌景观
  View of a clear, peaceful and natural canal
5 双标高步行系统立体多变的景观空间
  Two-level pedestrian system with varied multi-level landscape spaces
6 灵动、流畅的景观序列
  Flexible and flowing landscape sequence
7 灵动多变的特色灯柱
  Featured bollards

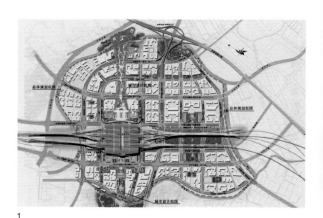

1

2

3

4

5

6

7

Landscape Design 景观设计

珠海玲玎海岸园林景观及绿化设计
Lingding Coastal Landscape and Greening Design, Zhuhai

项目位于珠海东澳岛上，是珠海市第一个海岛整体开发项目。全岛面积4.65平方公里，岸线长15.09公里，是万山群岛中的经典海岛。项目将建设成为国家4A级旅游景区，包括六星级的宾馆，集观光旅游、休闲度假、海上运动、会议培训、高级会所、高级运动球场为一体的多功能旅游景点，打造成世界一流的旅游度假胜地，现代的国际海岸，并成为珠海推广海岛游的示范岛屿。

项目设计上充分尊重海岛依山傍水的自然景观特色，保持并合理利用现状山地、沙滩、红石滩、相思林带等景观资源，形成地块特色明显的景观分区。设计结合项目休闲度假的"慢生活"定位，从人的心理和行为习惯出发，利用现状资源优势创造出丰富多样的景观活动空间，合理规划服务节点，为游人提供优质享受自然的场所。

在景观设计上，提取来自大自然的材料及与酒店风格统一的材料，使用纯洁的白、海洋蓝及石头、木条等天然材料的原色，营造自然和谐又舒适的度假天堂。在保留及修复原生植被前提下，根据现场地域特色筛选适合本地海岛种植的优良树种进行绿化配置设计，并形成特色鲜明的种植区域。核心景区南沙湾景区，适当增大沙滩面积，扩大沙滩容量，丰富沙滩设施和水上活动，打造出一个最具亚热带南国岛屿风光特色的旅游度假胜地景观。

Located at Dong'ao Island, Zhuhai, the Project is the very first whole-island development of Zhuhai. This 4.65 km² island with a coastline of 15.09 km long is a typical one among the Wanshan Archipelago The Project is planned as a national 4A tourist attraction with six-star hotels, integrating multiple functions such as sightseeing, leisure, marine sports, conference and training, high-end clubs and sports venues. It will be developed into a world-class tourism resort, modern international coast and an example for promoting island tourism in Zhuhai.

The designers fully respects the project's proximity to the natural landscape of the mountain and the sea. The existing landscape, such as the mountain, beach, red pebble beach and Acacia forest belt, are utilized for distinctive site zoning. With consideration of the project positioning of slow life and the psychology and behaviors of tourists, a variety of landscape activity spaces and reasonably planned service nodes are created based on the existing site resources, offering the tourists some quality places to enjoy the nature.

In landscape design, the materials extracted from the nature and those complying with the hotel style, coupled with the colors of pure white, ocean blue and those of the natural materials like stone and wood, jointly create a harmonious and comfortable holiday paradise. On the basis of preserving and restoring the native vegetation, fine tree species suitable for island are selected and cultivated for optimized greening design according to the local characteristics of the island, developing a planting area with distinct feature. The core scenic spot, Nansha Bay area, is expanded in beach area and capacity to offer more diverse beach facilities and water activities, completing the project with attractive landscape of the most typical resort on subtropical island in south China.

项目地点：珠海市万山区东澳岛
设计时间：2010.11
建设时间：2014.3
建设规模：约10万m²景观设计，3km绿道景观设计
曾获奖项：2015年度全国优秀工程勘察设计行业奖园林景观专项二等奖
2015年度广东省工程勘察设计奖园林景观专项一等奖
2014年中国风景园林学会"优秀园林绿化工程奖"金奖

Location: Dong'ao Island, Wanshan District, Zhuhai
Design: 2010.11
Construction: 2014.3
Size: Landscape design of 100,000 m² and landscape design of green roads of 3 km
Awards: The Second Prize of Garden & Landscape under National Excellent Engineering Exploration and Design Award, 2015
The First Prize of Garden & Landscape under Excellent Engineering Exploration and Design Award of Guangdong Province, 2015
Gold Prize of Excellent Landscaping Project Award of China Society of Landscape Architecture, 2014

1 总平面图
  Site plan

2 错落的景致和洁净的沙滩，热烈的海滩度假氛围
  Well-proportioned view and clean beach foster exciting atmosphere of beach resort

3 依山随势的台阶、平台、植物，营造步移景异的海岛风光
  Steps, terraces and plants are provided to follow the mountainous terrain, jointly presenting a fascinating island view that varies with the movement of the visitors

4 层层跌落、无边观海的浪漫泳池
  Cascading infinite pool with sea view

5 绿意围绕、海风轻拂、放松地SPA
  Relaxing SPA within a natural setting and see breeze

6 背山面海，层次丰富的园林
  Gardens nestling by the mountain and the sea

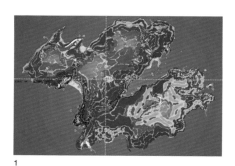

1

广州市金沙洲居住新城P线景观、K、M、N、U路绿化工程
Jin Sha Zhou Line P Landscaping and Roads K/M/N/U Greening Project, Guangzhou

金沙洲是广州市连接佛山的中心区，是广佛同城的重要枢纽。金沙洲P线滨江绿地是金沙洲新城的唯一滨江绿地及重要的生态景观廊道，在各市级滨江绿地中具有自己鲜明的特色，是金沙洲新城的一个地标性景观。

项目整体按带状公园设计，并依据现状的人文、自然资源条件，合理结合周边用地功能规划，以过江大桥为"界"，将该岸线绿地划分为艺术休闲区、文化休憩区、体育活动区，各区域之间有机连接，自然顺畅。南区段，以艺术休闲为主题，提出了保留南藤集团厂区部分建筑，并依据其"工艺制作"的内涵，将其改造成为以艺术创造为主的绿地活动场所；中区段，以文化休憩为主题，区内有历史保护建筑（西式别墅）、古庙、古树、石驳岸广场等，以此为契入点，设置了与人文景观气息浓郁的广场及休闲绿地；北区段，以体育活动为主题。呼应总体规划中关于此段设置体育运动场地的布局，结合该段用地较大的条件，设置篮球、网球、羽毛球等运动场所，并根据户外休闲运动的概念，进行体育公园设计及相关配套。项目整体设计新颖、美观，与周边环境协调统一。

沿着珠江支流白沙河西岸规划的长条绿带中以居民的步行尺度为依据在各区段内设置的景观节点、开敞空间关系明晰，较好地呼应了周边的用地功能性质，同时也满足滨江大道上交通视廊的需要。沿河岸的林荫步道上漫步，无论是白天或者夜晚沿河的城市景象都那么的迷人。项目建成后已然成为市民最喜欢去的休闲活动场所之一。

Jin Sha Zhou is a central area that connects Guangzhou with Foshan and an important hub for Guangzhou-Foshan integration. As the only riverside green space and a critical eco-landscape corridor of Jin Sha Zhou New City, the Line P riverside green space is a landmark of its own distinctive features in the area.

The Project is designed into a belt-shaped park. In view of the cultural and natural resources and the functional planning for the surrounding plots, the waterfront green spaces, which is bound by the river-crossing bridge, is divided into art and recreation area, culture and leisure area, and sports area. These areas are organically and naturally connected to each other. The south zone is themed on art and recreation. Some of Nanteng Group's factory buildings are preserved and transformed into green spaces for artistic creation in view of their former functions for process and manufacturing. The middle zone is themed culture and leisure. It accommodates historical buildings (western-style villas), ancient temples, time-honored trees and revetment square, based on which plazas and leisure green space are developed. The north zone in relatively larger area is themed on sports activities. It accommodates basketball, tennis, badminton and other sports venues as specified in the master plan, and sports parks and relevant supporting facilities based on the concept of outdoor recreational sports. The Project of novel and gorgeous overall design is harmonious with the surrounding environment.

Landscape nodes and open spaces placed in the longish green belt along the west bank of Baisha River, the tributary of the Pearl River, based on the walking scale of residents, are in clearly defined spatial relations, echoing with the functions of the surrounding plots and meeting the needs of visual corridor along the waterfront promenade. Roaming along the boulevards in daytime or nighttime, citizens can always enjoy the enchanting urban views along the river. The Project has become one of the most popular places for leisure activities since its completion.

项目地点：广州市白云区金沙洲
设计时间：2005.9
建设时间：2007.4
建设规模：总规模32.9万m²，道路绿化共9.13万m²，水岸长度约5km
曾获奖项：2011年度全国优秀工程勘察设计行业奖三等奖
2011年度广东省优秀工程勘察设计奖二等奖

Location: Jin Sha Zhou, Baiyun District, Guangzhou
Design: 2005.9
Construction: 2007.4
Size: 329,000 m²; road greening of 91,300 m²; riverside trail of 5 km
Awards: The Third Prize of National Excellent Engineering Design Award, 2011
The Second Prize of Excellent Engineering Design Award of Guangdong Province, 2011

1 总平面图
  Site plan
2 鸟瞰图
  Bird's eye view
3 开阔大气的广场与金沙洲大桥遥相对望
  Open and generous square facing Jin Sha Zhou Bridge in a distance
4 简洁的草坡地形营造清新的广场氛围
  Lawn of concise pattern fosters refreshing atmosphere of the square.
5 简洁秩序井然的户外舞台
  Simple yet tidy outdoor stage
6 繁花盛开的景观园路
  Landscaped parkway framed by blooming flowers
7 舒朗的树林草地，带来开阔的视野
  Large expanse of lawn with few trees offers open horizon

1

室内装修

GDAD室内装修设计，自20世纪80年代的钓鱼台国宾馆、中国工艺美术馆开始，到21世纪初的广州新白云国际机场、南方电网总部大楼、中山医肿瘤医院，再到现代的港珠澳大桥、广州地铁21号线、广州报业文化中心、南方航空总部大楼、南方医院惠侨楼、广州民俗博物馆、华为总部酒店（六星）等，在公共建筑领域以原创性、艺术性、唯一性不断创新发展壮大。

社会多元化高速发展，GDAD综合大院的优势，室内装修设计涵盖了城市综合体、总部办公、医疗养老、文化教育、商业广场、酒店建筑、交通建筑、体育建筑、城市改造等公共建筑。

公共建筑室内装修设计，以解决大众日常出行、购物、工作、健康、教育、休息社交、艺术鉴赏等日常必须为终极目标，让人于必经之余心情愉快、惊艳感叹、记忆深刻，提高人们的公共意识、提高审美情趣，影响大众的生活态度、出行习惯。公共建筑室内装修设计介于工程技术与艺术创作之间，既要满足工程技术的规范要求、规划要求、人流动线、材质应用、消防疏散、环保节能、工程造价等，又要突出其艺术创作的原创性、行业特征、空间个性、艺术感观，造型优美、视观惊艳，给人以愉悦感。

南方航空公司总部大楼，是一座大型集商业购物中心和甲级写字楼的综合体建筑。以国际化经营理念，体现企业文化、地域文化、国际视野和与企业影响力相匹配的形象感官，以"云端红棉"为主题，将蓝天白云、航空飞行、南航标识、广州红棉的元素进行提炼简化和升华，以现代设计手法创造国际化大企业形象的总部大楼办公空间：首层大堂以南航飞机标识形象作主壁形象墙，以蓝天白云喷绘装饰顶棚、以红棉点缀，以明显的企业文化、行业特征营造一个明亮宽敞的接待大堂；大办公区以内机舱形象提炼，创造优美弧度、灰白色舷窗的"空中"办公空间。以明显形象、个性感观让身处其中的人们倍感舒畅。

南方医院的惠侨楼是一座特殊的医疗大楼：中国最早的涉外医疗机构，接待多国元首和中国大多数国家领导人；它的医疗技术、建设规模、配套设施、装饰形象在国内80年代以来是最好的，最早将个性化服务、营养餐饮、舒适的住院环境、小众的医疗体验集中于一栋建筑物内。我们以现代的设计手法，再现其辉煌的历史，满足现代环境下特殊人群的高规范要求，续写其作为华南地区标杆式医疗机构形象。

GDAD室内装修设计作品荣获多次国际、国家级和行业设计奖，广州报业文化中心荣获加拿大GPD设计奖特别大奖、亚太设计奖金奖、中国室内装饰协会双年展金奖、台湾金点奖入围奖；港珠澳大桥东西人工岛荣获加拿大GPD设计奖铜奖和亚太设计奖铜奖；南方航空公司总部大楼荣获中国建筑学会铜奖和亚太设计奖优秀奖；广州天河文化艺术中心荣获亚太设计金奖；中山大学孙逸仙医院南院区荣获亚太设计奖金奖；中移动福建公司总部大楼荣获中国室内装饰协会双年展金奖。

Interior Design

GDAD's interior design team has seen continuous innovation and growth in the field of public buildings. The originality, artistry and uniqueness of our design has been witnessed by Diaoyutai State Guesthouse and China Museum of Arts and Crafts in the 1980s; Guangzhou Baiyun International Airport, the Headquarters Building of China Southern Power Grid, and Sun Yat-sen University Cancer Center at the beginning of this century; and Hong Kong-Zhuhai-Macao Bridge, Guangzhou Metro Line 21, Guangzhou Daily Group Culture Center, the Headquarters Building of China Southern Airlines, Huiqiao Building of Nanfang Hospital, Guangdong Folk Arts Museum, and Huawei Headquarters Hotel (6-star) at modern times.

Benefiting from GDAD's multidisciplinary background, its interior design service responds to the diversifying and fast-developing society and covers the public buildings like urban complex, headquarters offices, medical institutions and nursing homes, cultural and academic buildings, commercial plazas, hotels, transportation facilities, sports venues and urban renewal projects.

The ultimate purpose of interior design of public buildings is to address the issues related to the daily travel, shopping, working, health, education, social intercourse and artistic appreciation. The design should make people feel delighted, amazed and impressed, help improve their public awareness, aesthetic taste, and affect their attitude towards life and travel habits. Interior design of public buildings is a discipline between engineering technology and artistic creation. On one hand, it must meet the code requirements on engineering technology, planning. pedestrian circulation, material application, fire evacuation, environmental protection, energy conservation and project cost. On the other hand, it should highlight the originality of its artistic creation, industry characteristics, identity of space, artistic perception, aesthetic shape and stunning yet pleasant visual experience.

The Headquarters Building of China Southern Airlines is a complex integrating commercial shopping center and Grade-A office building. With modern design techniques, the space of the headquarters office building of an international company is created based on the international operation philosophy that reflects the corporate culture, regional culture, international vision and corporate influence and the theme of red kapok flower in the cloud that features the simplification and sublimation of various elements including the blue sky and white clouds, aviation, logo of China Southern Airlines and red kapok flower, Guangzhou's city flower: In the lobby on the first floor, the logo wall displays the logo of the China Southern Airlines, and the ceiling painted blue sky and white clouds is decorated with red kapok flower, creating a bright and spacious reception lobby featuring distinct corporate culture and industry characteristics; in large office area, the image of passenger compartments is referenced in creating an office with beautiful curves and gray porthole windows "in the air". The distinct and individualized ambience thus fostered make the people feel comfortable and delighted.

Huiqiao Building of Nanfang Hospital is a special medical building: it is China's earliest medical institution concerning foreign affairs, receiving the heads of state of many countries and most of China's leaders. It has been a leading domestic institution since 1980s in terms of medical technologies, construction size, supporting facilities, and interior decoration in China, and the first one to integrate personalized service, nourishing catering, comfortable hospital environment, and alternative medical experience in one building. With modern design techniques, we reproduce its glory and meet the demanding requirements of a certain patient group at the modern times, continuing to enhance its image as a benchmark medical institution in south China.

Our interior design works have won many international, national and industry design awards. Among them, Guangzhou Daily Group Culture Center won the Special Award of Grands Prix du Design Awards, Gold Award of Asia Pacific Interior Design Awards, Gold Award of the China International Interior Design Biennial Exhibition, and finalist of Golden Pin Design Award; the East and West Artificial Islands of Hong Kong-Zhuhai-Macao Bridge won the Bronze Award Grands Prix du Design Awards and the Bronze Award of Asia Pacific Interior Design Awards; the Headquarters Building of China Southern Airlines won the Bronze Award of the Architectural Society of China and Excellence Award of Asia Pacific Interior Design Awards; Guangzhou Tianhe Culture and Art Center won the Gold Award of Asia Pacific Interior Design Award; the South Campus of Sun Yat-sen Memorial Hospital, Sun Yat-sen University won the Gold Award of Asia Pacific Interior Design Award; the Headquarters Building of China Mobile Fujian Company won the Gold Award of the China International Interior Design Biennial Exhibition.

广州报业文化中心
Guangzhou Daily Group Culture Center

广州报业文化中心位于广州珠江畔，根据其良好的地理位置，结合广州地域文化，提出以"珠水流光"、"企业出彩"的设计理念，表达上通过灯光转化，以"线"的设计形式，通过不同色系的光，结合空间造型、软饰色彩在空间中做点缀，诠释设计主题。

本案在细部设计上，工艺大胆创新，触感精制顺滑，力求打造极具现代感、超凡脱俗的办公空间环境。整个设计，均采用柔和的灯光，和谐的色彩材质搭配，减少源头上灯光、色彩对人体视觉的刺激，更符合人们长时间舒适办公、人性化的需求。功能布局明确合理，在造型和光效上独具匠心，不仅满足各区域使用上的需要，更是从细微之处烘托出对人性的关怀，凸显超然的雅致与舒适。同时大量新型材料、新工法的运用将现代简约的办公空间梳理到极致。其种种运用前瞻性的语言，将自然要素与现代元素的有机结合，正是经典与时尚、优雅与品位、清新与自然的全新设计理念。

设计目标满足未来办公、生产及管理的功能需求、环境舒适优美、提高生产效率。设计要体现传媒企业特征，企业形象和企业文化的内涵，装饰风格要凸显个性，现代、美观和与周边建筑风格有机融合，达到"绿色环保，高效信息，人文生态"的要求，既体现高效节能，体现人文气息，体现企业对员工的关怀，又体现广州日报粤传媒作为大型传媒企业对新闻及社会的责任感。

项目地点：	广州市海珠区阅江西路
设计时间：	2013
建设时间：	2014
建筑面积：	地上建筑面积约12.6万m², 地下总建筑面积6.5万m²
建筑层数：	建筑T1、T2塔楼25层，其余高层建筑7层，地下3层
建筑高度：	156m
曾获奖项：	2015年加拿大GRANDS PRIX DU DESIGN 特别大奖 2013/2014年亚太室内设计精英邀请赛 金奖 2014年第十届中国国际室内设计双年展 金奖

Location:	Yue Jiang Xi Lu, Haizhu District, Guangzhou
Design:	2013
Construction:	2014
GFA:	1126,000m² aboveground and 65,000m² underground
Floors:	25 floors for T1 and T2, 7 floors aboveground and 3 floors underground for other highrises
Height:	156m
Awards:	GRANDS PRIX DU DESIGN Special Award, 2015 Asia Pacific Interior Design Awards For Elite-Gold Prize, 2013/2014 The Tenth International Interior Design Biennale of China Award-Gold Prize, 2014

Prominently located by the Pearl River in Guangzhou, the design concept of "flowing light by the Pearl River with fascinating appeal" is proposed by incorporating the regional cultural features. With the light translation and "linear" design, the colorful lights embellish the space in view of the spatial feature and color of soft decoration to interpret the design motif.

The detail design features bold and innovative process with sophisticated and smooth effect to create the office environment that is highly modern and incomparable. With the soft light and harmonious color mixture, the visual stimulation by the light and color is minimized from the source, allowing for long-time comfortable work with human-oriented design. The reasonable functional layout and ingenious feature and light effect not only caters for the demand of each zone but also shows the human care from details with unparalleled elegance and comfort. Meanwhile, a large number of new materials and construction methods are applied to streamline the modern and concise office space to the maximum. The forward-looking vocabulary and organic combination of natural and modern elements presents the new design concept that is classic yet trendy, elegant and tasteful, fresh and natural.

The design aims to meet the functional demand of future office, production and management with favorable environment and higher productivity. Apart from the manifestation of the features, image and culture connotation of modern media business, the decoration is individualized, modern, aesthetic and organically harmonious with surrounding building style to meet the requirement of being "green and environment-friendly, efficient and informationized, cultural and ecological". Reflecting the high efficiency and energy saving, cultural atmosphere and caring for the staff, it also shows responsibility of Guangzhou Daily as a large media business for the press and society.

2

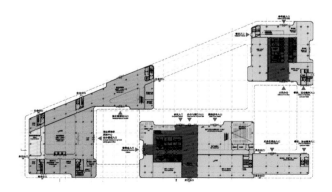

1 平面图
  plan

2 寓意"流光"的白色光带贯穿整个空间
  The white light belt implying flowing light penetrates through the whole space

3 根据不同的职阶由内而外分为四层，各层之间沟通合作紧密、信息传递迅速，为高效率的新闻工作提供了良好条件
  Four hierarchies are developed inside-out in view of the positions with close communication and cooperation and fast information transfer between hierarchies, ensuring the efficient work

4 游泳池通过光与顶棚的流动造型和颜色的变化，形象地将流光与红棉的意蕴表达得既动感又含蓄
The swimming pool by the flow of light and smallpox modelling and color changes, vividly conveys time, with the implication of red is both dynamic and implicitly

5 展厅为报业博物馆的展览馆的主要区域
The exhibition hall for the main area of newspaper of the museum's exhibition hall

6/7 立面图
elevation

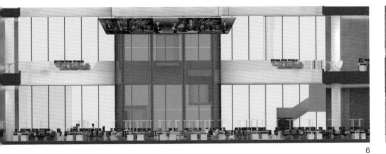

广州民俗博物馆
Guangdong Folk Arts Museum

广州民俗博物馆室内装修设计以时尚优雅的笔触极致的精确刻画，寥寥几笔就将空间勾勒得唯美灵动，"清新、自然、简约、优雅"这是广州市民俗博物馆的审美定位，更是以全球化设计理念，结合城市的历史文化、风貌特色、仿古特色，落定在该博物馆空间里，表现出最新的设计理念。

设计原则采用广府建筑装饰元素及风格，着重体现花都民俗和人文地域特色。坚持"可持续"设计方法，尊重现状且合理利用空间，配合建筑方案的构思，达到里外和谐统一。室内色彩依照博物馆的使用性质，整体色调明快，传统风格突出，同时利于展示布展。本案从项目定位出发，结合当地环境文化的尊重，提出设计主题"粤韵流芳"。主要空间以"越、粤、悦"作为主线元素贯穿，配合极富意境的表现章法，在抽象元素的基础上赋予设计独特的灵魂。元素从当地环境、自然条件、地域文化、民间艺术中巧妙提取，结合民俗博物馆的运行定位，以"粤"为目的，在室内设计中多角度，全方位地加以展示。

主色调传承中国笔墨山水画"黑白灰"搭配的灵气。粤语是广府文化最为突出的代表，而白纸黑字又是语言传承的最佳载体，同时提炼了资政大夫祠古建筑群中大量使用的灰色配色，将黑、白、灰以1:4:6的黄金比例进行配比，使得整个空间素雅而又纯净、宁静而柔和。

With trendy and elegant interior style, the aesthetic and flexible space is outlined precisely in a few strokes to portray the refreshing, natural, concise and elegant appearance of Guangdong Folk Arts Museum. The urban legacy, features and antique style are embedded in the museum space to reflect the up-to-date design concept.

By incorporating the decorative elements and style of Cantonese architecture, we highlight the local folk and cultural features of Huadu. Sustainable design approach, such as respect for the existing conditions and reasonable use of space, are employed to support the architectural design and realize the harmony and unity between the interior space and exterior appearance. In response to the functionality of the museum, the interior color is bright with highlighted traditional style to facilitate display and exhibition set-up. The design theme is of inheriting the Cantonese style is proposed in consideration of the project positioning and out of the respect for the local environment and culture. The leading elements of *Yue*, a homophone meaning excellence, Canton and joy all at the same time in Chinese, are embedded into the spaces, and, through the poetic expression, develop unique soul of the design on the basis of the abstract elements. These elements tactfully are extracted from the local environment, natural conditions, regional culture and folk art, then presented in an all-round way in the interior design to respond to the museum's operation positioning and the design theme of inheriting Cantonese style.

The main tone inherits the remarkable mixture of black, white and grey in the traditional Chinese ink landscape painting. The language of Cantonese is the typical representative of Cantonese culture, while the black characters on white paper are the best carrier for language inherence. The black, white and grey widely used in the ancient architecture Zizheng Dafu Hall are mixed at the golden ratio of 1:4:6, rendering the whole space neutral and pure, quiet and subdued.

项目地点: 广州市花都区新华街三华村资政大夫祠建筑群
设计时间: 2016
建设时间: 2017
建筑面积: 9000m²
建筑层数: 地面2层，地下1层

Location: Zizhen Dafu Hall, Sanhua Village, Xinhua Sub-district, Huadu District, Guangzhou
Design: 2016
Construction: 2017
GFA: 9,000m²
Floors: 2 aboveground and 1 underground

1 中庭设计简洁明快
  Concise and lively atrium

2 报告厅的设计整体而简洁
  Holistic and concise lecture hall

3 主入口外观延续了资政大夫祠入口的古典设计元素
  The appearance of the main entrance incorporates the classic entrance elements of Zizheng Dafu Hall

4 平面图
  Plan

5 大堂空间主色调传承中国笔墨山水画"黑白灰"搭配的灵气
  The main tone of the lobby inherits the remarkable mixture of black, white and grey in Chinese ink landscape painting

1

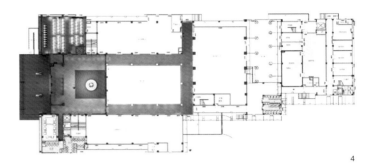

中国南方航空大厦室内装饰装修工程设计
Interior Fit-out for China Southern Airlines Building, Guangzhou

在白云之端,有着人类对天空的无尽向往及遐想;在大地之南,有着英雄红棉对一方水土一方人的指导及感染。本案设计以"云端红棉"为主题,以"云端"中的行云光线元素,"红棉"意念抽象化元素,以现代的艺术审美、设计手法、施工工艺创造一个现代、时尚、简约而又具有浓厚文化氛围的办公空间。

设计主色调以经典的黑白灰配色,飘逸弧线造型(墙体、顶棚)、弧形灯光走向,弧形家具摆布,寓意"云端"的行云光线贯穿整个室内空间;"红棉"意念在细节中自然铺开:红色的沙发、红色的灯座、红棉挂画、红棉标识贯穿始终,营造国际化企业总部大楼所应有的、所感受到的高端、大气、宽阔、时尚。同时将南方航空总公司(全球第三大航空公司)的行业特点、地域文化、企业文化以及全球化经营理念以视觉感受淋漓尽致地表现出来:航空、南方、奋发、包容。

材料上采用环保而耐用的石材、金属板、玻璃以轻工业风的工艺制作安装,强调低耗、节约、重复使用的设计原则。灰色不锈钢面板搭配起来时尚而动感。"灰"、"白"色调的简单搭配,使得整个空间素雅而又纯净、宁静而柔和。二者结合,表现出如同坚实流畅起伏的如英雄一般的肌肉线条,在优雅的灯光下动荡起伏,随着视角变化,在阳光和灯光的不同照射下散发十足的动感力量。

项目地点:广州市白云区白云新城云城东路地段AB2910002地块
设计时间:2013
建设时间:2014
建筑面积:建筑-4至36层,装修室内建筑面积约11万m²
建筑层数:首至6层为裙楼;地上总计36层,4层地下室
建筑高度:150m
曾获奖项:2014/2015年亚太室内设计精英邀请赛 优秀奖
2015年中国建筑学会第十八届中国室内设计大奖赛 铜奖

Location: Plot AB2910002, Yun Cheng Dong Lu, Baiyun New Town, Baiyun District, Guangzhou
Design: 2013
Construction: 2014
GFA: B4 to F36 with an interior fit-out area of 110,000m²
Floors: F1-F6 for podium, 36 aboveground and 4 underground
Height: 150m
Awards: Elite-Excellence Award of Asia Pacific Interior Design Awards, 2014/2015
Bronze Prize of the Eighteenth Interior Design Award by the Architectural Society of China, 2015

The design theme is kapok on clouds, with the white cloud symbolizing people's aspiration about the sky and the kapok tree (known as the hero tree) the spirit of peoples in Guangzhou. Design elements drawn on from sunlight rays over cloud and red kapok flower are used to create modern, trendy, concise and highly cultural office spaces through the modern artistic aesthetics, design approach and construction techniques.

The tone is dominated by the classic mixture of black, white and gray. The concept of sunlight rays over clouds is woven into the whole interior space through the graceful curved shape (wall and ceiling), arc lighting and cambered furnishing, while the image of red kapok flower is naturally reflected through details, such as the red sofa, lampholder, kapok painting and logo, which altogether presents a high-end, lofty, spacious and trendy atmosphere tailored for an international business headquarters building. Meanwhile, the industrial feature, local culture, corporate culture and global business philosophy of China Southern Airlines (the world's third largest airline) are fully visualized i.e. aviation, southern, forging ahead and inclusiveness.

The environment-friendly and durable stone, metal plate and glass are fabricated and installed in a light industrial style to emphasize the design principle of low energy consumption, conservation and recycling. The grey stainless steel panels appear trendy and dynamic, while the concise mixture of the grey and white makes the whole space tasteful, pure, quiet and mild. Such a combination portrays the sweeping masculine lines that undulate dynamically in sunlight and lighting along with the change of viewing angles.

1 首层大堂平面图
  Floor plan of F1 lobby

2/6 大堂空间凸显企业特色
  Distinctive corporate features in lobby

3 空间围绕核心筒而设,设计通过天花造型强调空间动线
  Spaces center on the core with the circulation underlined by the ceiling pattern

4 办公室简约时尚
  Concise and trendy office

5 花瓣造型顶棚让空间更显大气,就餐时增加食欲
  The petal-shaped ceiling creates lofty space and a pleasant dining environment.

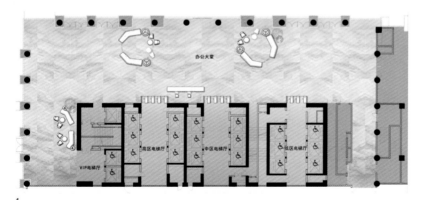

1

2

3 4 5

6

南方医院惠侨楼室内装修工程设计
Interior Fit-out for Huiqiao Building, Nanfang Hospital, Guangzhou

本案设计结合了南方医院的文化特色，提出了"生命之树，健康呵护"的设计理念。采用了现代时尚大气的设计手法，意欲创造出一个具有亲和力、温馨感的现代医疗空间。为病人提供一个舒适安静的环境，同时也改善医疗工作人员的办公条件，为病人提供更好的服务。

本案设计以"树"作为设计主线。通过对"树"的各种形态进行分析研究，提炼出具有现代设计感、抽象的装饰元素，巧妙地运用到每一个空间里面，营造出一种大自然的氛围，让病人时刻置身于"大自然"当中，亲近大自然，亲吻大自然。同时也呼应了"生命之树，健康呵护"的设计理念。材料和色彩搭配上较为温和。材料上选用大自然常用的木饰面、铝扣板、烤漆玻璃、GRD材料、pvc地板胶、人造大理石、防滑砖。色彩搭配上：白色、木色、绿色作为主色，体现了时代感和温馨感，使病人有"家"的亲切感，同时也呼应设计主题。

空间整体以柔和的温馨色调为主，给病人温暖的感受。我们的设计风格摆脱了以往"白色恐怖"的医院氛围，意在营造温馨之家。顶棚、地面是白色，立面我们运用了木色，木色使人亲近，这种搭配一方面体现出医院的洁净度，又能给人温暖的感受。重要空间的顶棚运用了镂空造型，丰富了空间层次。

The design concept, i.e. the tree of life for health and care, incorporates the cultural features of Nanfang Hospital. The modern, trendy and noble design approach aims to create the amicable and warm modern medical treatment space to provide the patient comfortable and quiet environment and improve the working conditions of the medical staff to better serve the patients.

With the tree as the design theme, the abstract decorative elements with strong sense of modern design are extracted based on the analysis and study for various forms of tree and skillfully integrated into each zone to create a natural ambience. This way, the patients are indulged into the nature in support of the design concept. The materials and color mixture are mild. The materials include wood veneer, aluminum gusset, paint glass, GRD, PVC flooring, artificial marble and non-slip floor tiles. In terms of the color mixture, the white, wood and green dominate to communicate the modern trend and warm feeling, making patients feel like at home and echoing the design theme.

The overall space is dominated by the soft and warm tone to bring about a cozy atmosphere. Instead of the chilling white of traditional hospital atmosphere, we aim to create a cozy home. With white ceiling and floor, walling in friendly wood color contributes to a clean and cozy environment for the patients The hollow-out ceiling at main spaces diversifies the spatial hierarchy.

项目地点：南方医科大学南方医院
设计时间：2014.12
建设时间：2015.7
建筑面积：20000m²
建筑层数：15
建筑高度：75m

Location: Nanfang Hospital of Southern Medical University
Design: 2014.12
Construction: 2015.7
GFA: 20,000m²
Floors: 15
Height: 75m

1 首层平面图
  F1 plan
2 共享阳光厅
  Shared sunlit hall
3 简洁温馨的休息厅
  Concise and warm lounge
4 岭南中式风格装饰，氛围庄严、奢华
  The Chinese-style decoration with Lingnan features creates solemn and sumptuous atmosphere
5 以木棉花和树为设计理念的大堂设计
  Lobby design themed on kapok flower and tree
6 大堂立面图
  Lobby elevation

1

2

3

4

5

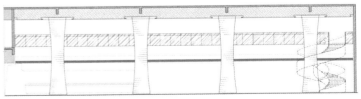

6

Classic Works 经典作品

北京钓鱼台国宾馆12号楼
Beijing Diaoyutai State Guesthouse, Building 12#

建设时间: 1983
Construction: 1983

建筑是按接待外国元首标准设计的高级宾馆，有贵宾休息厅，大会客厅，会议厅，宴会厅，元首级豪华套间及休息内庭等。建筑面积4400平方米，于1981年设计，1983年竣工。建筑造型层次丰富，屋面采用玻璃琉璃瓦及双脊瓦，勾头滴水等。宾馆内设有较先进的现代化设备。该宾馆建成使用后曾多次受到外宾及领导赞扬。该工程于1984年获国家优秀设计金质奖。1984年获广东省优秀工程设计一等奖。

The building is a high-end guesthouse for receiving foreign heads of state. Designed in 1981 and completed in 1983 with a GFA of 4,400m², the Guesthouse includes VIP lounges, large reception halls, meeting halls, banquet halls, state head luxury suites and rest chambers, which are supported by sophisticated modern amenities. It features hierarchical architectural form, with glazed tiles, double ridge tiles and eave tiles being used on the roof. Highly praised by foreign guests and leaders, the project won the Gold Award of National Excellent Design and the First Prize of Excellent Engineering Design of Guangdong Province in 1984.

深圳国际金融大厦
Shenzhen International Finance Building

建设时间: 1983
Construction: 1983

深圳国际金融大厦是一座全幕墙超高层塔式建筑，坐落于繁华的罗湖区建设路与迎春路交汇处。大厦占地4500平方米，总建筑面积52300平方米。大厦高38层，134.5米。大厦设计布局灵活，分区明确，基本平面由二个矩形组合；在近乎对称的大矩形中突出全对称之矩形主体建筑。矩形标准层平面，可以分层作为一个大单元出租，也可以分割为2~4个单元出租。大厦主体结构为矩形塔式建筑。由中筒及四角钢筋混凝土筒体组合。中筒为大厦之竖向交通和设备管井。

Located at the intersection of Jian She Lu and Ying Chun Lu in bustling Luohu District, Shenzhen International Finance Building is a fully-glazed super high-rise tower. Occupying a land area of 4,500m², the 134.5m building has a GFA of 52,300m² and 38 floors. With flexible layout and clear zoning, the building features two rectangles in plan. The fully symmetrical rectangular main building stands out of the nearly symmetrical larger rectangle. The rectangular typical plan can be leased out in whole or divided into 2 to 4 units. This rectangular tower is made of central core housing vertical transportation facility and MEP shafts and reinforced concrete frame-core at four corners.

广州国际科技贸易展览交流中心(中央酒店)
Guangzhou International Technology Trade Exhibition and Exchange Center (Central Hotel)

建设时间: 1986
Construction: 1986

广州国际科技贸易展览交流中心,是一座为科技展览贸易交往活动而兴建的大型、综合性的建筑。包括国际展览馆,出租办公楼和中央酒店三大部分,有展览厅6300平方米,出租办公楼4000多平方米,旅业额房234套和400多座的多功能会议厅,总建筑面积4万多m^2,楼高6层,各种设施设备齐全、先进。该工程于1986年3月建成,1988年获广东省优秀设计一等奖,1989年获建设部优秀设计二等奖。

Guangzhou International Technology Trade Exhibition and Exchange Center is a large mix-used building for the technology exhibition and trade. It includes three parts, i.e. the 6,300m^2 exhibition hall, 4,000$^+$m^2 leasable office and central hotel with 234 guestrooms and a 400-seat multi-purpose conference room. The full-fledged building is designed with a GFA of 40,000$^+$m^2 and 6 floors. Completed in Mar 1986, the Center won the First Prize of Excellent Design Award of Guangdong Province in 1988 and the Second Prize of Excellent Design Award of the Ministry of Construction in 1989.

北京大学理科教学楼群
Science Teaching Building Complex of Peking University

建设时间: 1986
Construction: 1986

教学楼群坐落在著名的北京大学校园内,占地面积9公顷,建筑面积11.4万平方米,包括十个系,两个研究所、两个馆、两个中心和一个公共教室区。该工程是国家"七五"计划的重点建筑项目之一,该设计方案经全国性投标并一举中标。本楼群设计充分考虑了它的特定环境,使其在校园中能与原有建筑群,原有环境取得协调,同时还要体现新时代的建筑风采。

The building complex is located inside the famous Peking University, with a land area of 9 hectares and a GFA of 114,000m^2. It accommodates ten departments, two research institutes, two halls, two centers and one public classroom. GDAD won the nationwide design competition of this important project listed in the national 7th Five-Year Plan. The design gives full consideration to the specific environment of the complex, so as to harmonize it with the existing campus buildings and environment while showcasing the architectural image in the new era.

北京陶然宾馆
Beijing Taoran Hotel

建设时间: 1986
Construction: 1986

位于北京市东城区（原崇文区）二郎庙，永定门火车站的东侧，太平街后的西侧，地理位置优越，环境优美。该工程以会议接待为主，亦同时对外经营为400间标准客房，其他配套设施齐全。总建筑面积3万平方米，地上22层，地下2层，建筑造型以多变和高低错落的屋檐，配以孔雀蓝的瓦面，线灰白色的墙面，力求与附近环境配调，体现民族特色。

The hotel is situated at Erlang Temple in Chongwen District of Beijing. It sits on the east of Yongdingmen Railway Station and the west of Tai Ping Jie, enjoying superior geographic location and pleasant environment. The Project is planned mainly for meetings and reception, meanwhile has 400 typical guest rooms and complete supporting facilities for external guests. With a total floor area of 30,000m², 22 aboveground floors and 2 underground floors, the building features changeful and staggered eaves, peacock blue tiling and white plastering walls, perfectly harmonizing with the surrounding environment and reflecting the national characteristics.

深圳图书馆
Shenzhen Library

建设时间: 1986
Construction: 1986

位于上步区荔枝公园西北角，占地2.4公顷，是深圳特区八大文化建设之一。工程总体布局依据地形地势，把建筑物作迭级处理，错落有序，在建筑造型上有民族特色，沿湖边布置的阅览室墙体挑出，与环境相融合。正门上方用倾斜的镜面玻璃，扩大了空间景观。全馆建筑面积14000平方米，包括有大厅、目录厅、报告厅、馆内设有中文报刊阅览，社科阅览，自然科学阅览，少年儿童阅览港台图书馆阅览等各种阅览室，共有一千多个阅览座位，各有开架和半开架，总藏书量63万册。

Located at the northwest corner of Lychee Park in Shangbu District, the Library is one of the eight major cultural projects in Shenzhen and occupies a land area of 2.4ha. The well-proportioned building with national features cascades to follow the topography. The reading room along the lake is designed with the cantilevered wall, well blending into the environment. The inclined mirror above the main entrance broadens the spatial landscape. The 14,000m² library of 630,000 books consists of lobbies, catalogue halls, lecture halls, as well as various reading rooms with a total of over 1,000 seats, offering Chinese newspapers and periodicals, social science books, natural science books, children's books, and books published in Hong Kong and Taiwan in open and semi-open shelf.

广东大厦
Guangdong Hotel

建设时间: 1987
Construction: 1987

东莞理工大学教学综合楼
Teaching Complex of Dongguan University of Technology

建设时间: 1988
Construction: 1988

广东大厦是广东省人民政府的会议宾馆，也是第六届全国运动会的配套工程项目之一。功能、设备齐全。总建筑面积58000平方米。地面19层，地下二层，内设516套标准客房，400座多功能会议中心和12个中小会议室，宴会厅和大小餐厅13个，约1400座咖啡厅等，还设有商场、洗衣房，可停泊150辆小汽车的地下车库和职工用房等。大厦于1987年10月建成。1988年获广东省优秀设计一等奖，1989年获建设部优秀设计三等奖。

As the hotel hosting the meetings for the People's Government of Guangdong Province, Guangdong Hotel is one of the projects built to support the 6th National Games. The 58,000m² hotel is full-fledged with 19 floors aboveground, 2 floors underground, 516 typical guestrooms, a 400-seat multi-purpose conference center, 12 medium and small meeting rooms, 13 ballrooms and restaurants of varied sizes, a 1,400-seat café, shops, laundry, a 150-space underground garage and staff rooms. Completed in Oct 1987, the hotel won the First Prize of Excellent Design Award of Guangdong Province in 1988 and the Third Prize of Excellent Design Award of the Ministry of Construction in 1989.

校园设计充分利用地形，主体建筑教学楼建筑面积46000平方米，东西长500米，南北宽400米，平面采用7个"口"字形组合，向左右逐级伸展，形成一级轴线明确的建筑群体，建筑体型层层跌落，首层以支柱层贯通并布置庭园绿化空间流通，层次丰富，立面为"口"字形方窗处理，虚实对比，富有时代气息。该工程于1990年4月建成。1993年获广东省优秀建筑设计二等奖。

The campus design makes full of the terrain. The main building measuring 500m long from east to west and 400m wide from south to north totals a floor area of 46,000m². Its plan is composed of 7 rectangle shapes extending to the left and right respectively into a building cluster with a clearly defined primary axis. The building shapes decline by floor until the pillared ground floor where a multi-level landscaped courtyard for circulation is provided. The façade employs rectangular windows to present a contrast between the solidness and void and the spirit of the times. The project was completed in April 1990 and won the Second Prize of Excellent Design Award of Guangdong Province in 1993.

北京中国工艺美术馆
China National Crafts & Arts Museum, Beijing

建设时间: 1989
Construction: 1989

中国工艺美术馆位于北京西长安街、复兴门立交桥的东北角,占地10,300平方米,建筑面积43,000平方米。建筑布局采用了步步深入、引人入胜的传统手法。主体建筑由序厅、展览、办公三部分组成。中部以序厅为中心,作为展厅的序幕并为展览以及参观的主要出入口。一二层为工艺品购物中心,三、四层为展览场地,五层为陈列馆。该工程1991年获建设部优秀设计三等奖,1991年获广东省优秀设计一等奖。

Located on Xi Chang An Jie, at northeast corner of Fu Xing Men viaduct, the 43,000 m² China National Crafts & Arts Museum occupies a land area of 10,300 m². With an engaging layout created out of traditional approach, the main building is composed of three parts, i.e. preface hall, exhibition hall and offices. The preface hall in the middle serves as the prelude to the exhibition halls and the main access to visit circulation. F1 and F2 are designed for artware shops, F3 and F4 for exhibition and F5 for display hall. The project won the Third Prize of Excellent Design Award of the Ministry of Construction and the First Prize of Excellent Design Award of Guangdong Province in 1991.

广东国际大厦63层
Guangdong International Hotel

建设时间: 1990
Construction: 1990

广东国际大厦坐落于广州市环市东路商业中心区,占地19540平方米,总建筑面积178000平方米,是由一幢63层主楼、两幢30多层副楼与五层群楼组成的,以金融为中心的多功能综合性建筑群体,总高200.18米。主楼平面采用了变化的八角形,立面既非笔直而又挺拔向上,这就是建筑造型上模棱手法的效果。立面上配以银色的蜂窝铝板外墙,浅蓝色玻璃和浅紫色的横向窗间墙,色调轻巧明快,更加强了主楼挺拔度和时代感。大厦主楼为塔式筒中筒钢筋混凝土结构,内外筒之间的楼板采用无粘结预应力平板结构,在高层建筑应用无粘结预应力楼板和地震区高层建筑采用钢筋混凝土结构方面达到国际先进水平。

Guangdong International Hotel is located in the commercial core of Huan Shi Dong Lu, Guangzhou, occupying a land area of 19,540 m². With a GFA of 178,000 m² and a total height of 200.18m, this finance-centered multifunctional complex is composed of one 63-floor main building, two 30-floor or so annex buildings, and one 5-floor podium. The main building is designed with a changing octagonal plan and a non-straight yet lofty façade, by virtue of edges and corners approach for architectural shape. The façade features external wall made of silvery aluminum honeycomb boards, light blue glazing and light purple horizontal pier between windows. The bright colors further enhance the loftiness and contemporary sense of the main building. The main building employs tower-type core-in-core reinforced concrete structure, with unbounded pre-stressing slab construction between the inner and outer cores. It takes the lead around the world in applying unbounded pre-stressing floor slab to a high-rise building and adopting reinforced concrete structure in a high-rise building in seismic region.

珠海市体育中心
Zhuhai Sports Center

建设时间：1990
Construction: 1990

珠海市体育中心位于珠海新香洲区的西面，该中心总用地面积42.2公顷，拥有40000个座位的体育场，5500个座位的体育馆，1500个座位的游泳馆，同时还有训练馆、网球中心、射击馆及各种训练场地，各场馆全部按照国际标准设计，不仅能举办各种大型国际比赛，还可作为大型文化娱乐的场所，供市民游乐和休息，成为适合人民参与活动的大型体育公园。该体育馆网壳设计获中国钢结构协会、空间结构协会"第一届空间结构优秀工程设计三等奖"；体育馆主体、屋架网架结构获中国建筑学会的"建筑结构优秀设计三等奖"；该工程2001年获"广东省优秀工程设计一等奖"。

Located in the west of Xiangzhou District, Zhuhai City, the Sports Center comprises the 40,000-seat stadium, 5,500-seat gymnasium, 1,500-seat natatorium, training hall, tennis center, shooting hall and various training fields on a land of 42.2ha. All facilities are designed up to the international standard to host major international events. It is also a large sports park to entertain the citizens with cultural and entertainment activities. The latticed shell of the gymnasium won the Third Prize of the first session of Structure Excellent Project Design Award by Association for Spatial Structures, China Steel Construction Society; the main structure, roof truss and grid structure of the gymnasium won the Third Prize of Building Structure Excellent Design Award by the Architectural Society of China; the Project won the First Prize of Excellent Engineering Design Award of Guangdong Province in 2001.

深圳香格里拉大酒店
Shangri-La Hotel, Shenzhen

建设时间：1991
Construction: 1991

深圳香格里拉大酒店是香格里拉国际集团属下廿多家豪华酒店的新成员，总建筑面积为64000平方米，高114.1米。酒店坐落在深圳火车站站前广场北侧，罗湖区南北两条主要干道：人民南路和建设路交汇处的繁华地带，位置优越、交通方便。酒店与火车站等其他建筑共同组成了广场建筑，借用了广场的空间和绿化，美化了环境和城市面貌。酒店占地面积8170平方米，建筑用地面积与建筑面积比值达到1：7.9。除主入口留有必要的停车场地外，建筑外墙基本沿建筑红线布置。

As the newcomer to the over-20-member luxury hotel family of Shangri-La Group, the 64,000m^2 and 114.1m hotel is prominently located in the north of Shenzhen Railway Station Square, the busy intersection of Ren Min Nan Lu and Jian Shen Lu, the two trunks of Luohu District with high accessibility. The hotel, railway station, other surrounding buildings and the spacious greened square jointly constitute an attractive city image. Occupying a land area of 8,170m^2, the hotel reaches 1:7.9 in the ratio between land area and floor area. Except for the necessary parking lot at the main entrance, the external wall is basically planned along the building property line.

广州国际贸易中心大厦
Guangzhou International Trade Center

建设时间: 1992
Construction: 1992

广州国际贸易中心大厦（潮汕大厦）位于广州天河北路与林和西路交汇的西北角，占地6032平方米，总建筑面积6.8万平方米，地上42层，地下2层，是一幢集展销、商住、办公于一体的多功能综合大楼，是设备先进与齐全、处于地理位置显要、交通方便的高科技智能大厦，总高166.8米。1998年大厦主体结构、建筑结构获中国建筑学会优秀设计一等奖。1999年获广东省优秀设计一等奖。

Guangzhou International Trade Center (Chaoshan Building) is located at the northwest corner of the intersection of Tian He Bei Lu and Lin He Xi Lu, with a land area of 6,032 m^2, a GFA of 68,000 m^2 contributed by 42 aboveground floors and 2 basement floors, and a total height of 166.8m. It is a multifunctional complex integrating sales exhibition, commercial residence and office, and a self-contained, prominently located and highly accessible hi-tech smart building. Its main structure and building structure won the First Prize of Excellent Design Award of Architectural Society of China in 1998 and the First Prize of Excellent Design Award of Guangdong Province in 1999.

汕头市委办公楼
Office Building of Shantou Municipal Party Committee

建设时间: 1994
Construction: 1994

位于汕头市海滨路北侧，东侧迎宾馆，南临汕头港湾。与岩石风景区相望，景色优美，是一幢行政办公综合楼，用地面积15660平方米，总建筑面积2.15万平方米，共10层，建成后获广东省1995年优秀设计一等奖，建设部二等奖，国家优秀设计铜质奖。

Located in the north of Hai Bin Lu, the office building neighbors Shantou Guesthouse on the east and Shantou harbor on the south, looking across the beautiful Rock Scenic Spot. This 21,500m^2 and 10-floor administrative office complex occupies a land area of 15,660m^2. The project won the First Prize of Excellent Design Award of Guangdong Province in 1995, the Second Prize of Excellent Design Award of the Ministry of Construction and National Bronze Award for Excellent Engineering Design.

广州东骏广场
Guangzhou Dongjun Plaza

建设时间: 1995
Construction: 1995

位于广州市东风东路与梅东路交汇之西南角，四面环路，主面朝北。落成于1995年10月，用地面积8500平方米，总建筑面积96900平方米，地上36层，地下2层，楼高133米，是一栋设计独特、装潢典雅的商业楼宇。流畅的平面曲线使立面造型飘逸，为了增加建筑形体的流动感觉，设计中采用逐段加高的处理手法，让建筑主体从30层至36层呈台阶状逐渐提高，似一面高高飘扬的旗帜，在蓝天白云的映衬下，倍添风采。

Located at the southeast corner of the intersection of Dong Feng Dong Lu and Mei Dong Lu, the road-embraced plaza has its main address on the north. Completed in Oct 1995, this distinctive and elegant commercial building occupies a land area of 8,500m² with a GFA of 96,900m², 39 floors aboveground, 2 floors underground and building height of 133m. The smooth planar curve develops elegant façade. To realize a flowing form, the building is escalated from F30 to F36 in a stepped way, resembling a flying flag against the blue sky.

东莞银城大厦
Dongguan Yincheng Building

建设时间: 1995
Construction: 1995

东莞银城大厦位于东莞市篁村，广深高速公路旁，交通方便，用地面积12000平方米，总建筑面积4.74万平方米，地上29层，地下一层，建筑总高度99米，是以按现代化星级酒店设施设计为主和银行营业办公用房为附组成的综合性建筑，建筑造型分上、中、下三段，下段为基座，强调突出四条柱，中段由正方形，每隔四层缩级扭转变菱形，上段配合结构筒体以尖塔的造型，使建筑物高耸向上。1997年获广东省优秀设计一等奖。

Dongguan Yincheng Building is prominently located at Huang Cun Village of Dongguan City by Guangzhou-Shenzhen Expressway. Planned with a site area of 12,000 m² and a GFA of 47,400 m², the building has 29 aboveground floors and 1 underground floor that total a building height of 99m. It is a complex designed mainly as modern star hotel but also with some banking office facilities. The building is divided into three sections, i.e. the upper, middle and lower section. The lower section is the base where four columns are highlighted; the middle section in square geometry sets back and twist into diamond every four floors; the upper section is a combination of a spire and the structural core, enhancing the verticality of building. The building won the First Prize of Excellent Design Award of Guangdong Province in 1997.

Epilogue 后记

陈雄

GDAD副院长、总建筑师
全国工程勘察设计大师
教授级高级建筑师
国家一级注册建筑师

今年是建院65周年，是省院发展历程上一个非常重要的关键节点。因为我们不仅面临国家社会经济的重大转型，而且面临建筑行业的重大变革，同时还面临省院改制建企，如何承前启后，适应新常态，创新求发展，于省院是挑战与机遇并存。

在这个重要的时刻，我们需要一个回顾、总结和展望的机会，因此去年院里决定重新编辑出版作品集。作为以设计为主业的技术服务机构，持守技术本源，致力设计创新，无疑是永恒的主题。设计作品是企业品牌的重要载体。省院的特点是建院时间长，专业比较齐全，技术积累深厚。新的作品集首次以各类技术成果综合展示省院的实力。同时，与以往相比，这次的编辑选择近年新完成或正在实施的作品，并且更加突出了作品集的可读性，以图片及说明充分表达作品的设计理念、技术创新及设计特色。

今天，当我们翻开这本作品集的时候，全院员工辛勤劳动、艰辛付出和努力拼搏的成果跃然纸上！省院近年来为社会为业界又贡献了一批富有影响力值得全院骄傲的建筑作品及技术成果。此时此刻，我们大家都应该为能够取得这些成果感到自豪！借此机会，衷心感谢各界朋友多年来的信任、支持和厚爱！没有你们的大力帮助，我们不可能有机会实现这些作品，不可能有机会取得技术进步。

在这里我们要特别感谢广东省建筑设计研究院党委曾宪川书记为本作品集作序。特别感谢我院顾问总工程师、全国工程勘察设计大师陈宗弼总工为本作品题写封面。特别感谢江刚、洪卫、孙礼军等主编一起讨论编辑方针，选择分类项目，修改完善排版。

感谢各位总工撰写各类建筑及技术成果的总结，他们分别是文化会展建筑（江刚）、科技教育建筑（梁彦彬）、商业、办公及酒店建筑（孙礼军）、交通建筑（陈雄）、医疗建筑（黄佳）、体育建筑（潘勇）、地下空间（洪卫）、居住建筑（周文）、规划设计（李鹏）、市政桥道（陈伟）、市政给排水（李骏飞）、水处理与环保技术（陈伟雄）、环境保护（原效凯）、景观设计（郭奕辉）、室内装修（冯文成）。

感谢各相关项目负责人组织整理了项目材料，感谢各相关部门给予的大力支持。

尤其要感谢院技术质量管理部的统筹组织编撰工作，蔡晓宝、郭伟杰和丁漫原几位同事不辞劳苦，做了大量具体工作。

廖荣辉先生为本作品集排版。他不辞劳苦，反复修改调整，追求完美的专业精神令我们十分感动。感谢叶飚先生为我们拍摄了部分作品的照片。感谢梁玲团队为作品集所做的全部翻译工作。

本作品集还引用了其他摄影师为部分项目拍摄的相关照片，在此一并表示衷心感谢！

Chen Xiong

Vice President, Chief Architect, GDAD
National Engineering Survey and Design Master
Professorial Senior Architect
Grade I Registered Architect

The 65th anniversary of GDAD this year marks an important milestone in its history. Facing the profound social-economic transformation of the country and the significant changes in the building industry, as well as GDAD's institutional reform, we must learn from the past, adapt to the new normal and innovate for development. While moving into the future, we are embracing not only the challenges but also opportunities.

At this critical moment, we need an opportunity to review and summarize the past and look into the future. So we decided to re-edit and publish this Collection. As a design-focused technical service provider, we have been unwaveringly developing our technical competence and promoting the design innovation, making each of our project a showcase of our brand name. GDAD has been renowned for its long-standing history, fully-fledged disciplines and profound technical know-how. For the first time, the new Collection presents our expertise and competence through the various types of design and technical accomplishments. Compared with the previous version, the new Collection includes more newly completed or ongoing projects in recent years, and devotes more attention to its readability which is achieved through photos and narratives that clearly elaborate the design concepts, technological innovations and design features of the projects.

The Collection is a loyal witness to the great diligence and efforts of all GDAD people. In recent years, we have completed dozens of influential projects and made impressive technical accomplishments that we are all proud of. At this moment, we are all proud members of GDAD! Here we would like to extend our sincere gratitude to all our friends for their trusts and supports over the years. Without their generous supports, it would be impossible for us to implement these projects nor made the technical advancement.

Our special thanks go to our Party Secretary Zeng Xianchuan for writing the Foreword to the Collection, our advisor Chief Engineer and National Engineering Survey and Design Master Chen Zongbi for handwriting the book title on the cover page, and editors-in-chief Jiang Gang, Hong Wei, Sun Lijunand etc. for establishing the editorial guidelines, defining the project categories and refining the typesetting.

We would like to thank the chief engineers for their efforts in summarizing the architectural and technical works in different sectors including cultural (Jiang Gang), S&T and academic (Liang Yanbin), commerce, office and hospitality (Sun Lijun), transportation (Chen Xiong), healthcare (Huang Jia), sports (Pan Yong), underground space (Hong Wei), residential (Zhou Wen), planning (Li Peng), bridge and road (Chen Wei), municipal water supply & drainage (Li Junfei), water treatment and environmental protection projects (Chen Weixiong), environmental protection (Yuan Xiaokai), landscape (Guo Yihui), , and interior (Feng Wencheng).

We appreciate the efforts of the project principals in gathering the project data and information and the generous support from all relevant departments.

In particular, we would like to thank Cai Xiaobao, Guo Weijie and Ding Manyuan from our Technical Quality Management Dept for their strenuous efforts in handling numerous works in great detail.

Our thanks also go to Liao Ronghui who has impressed us with his professionalism in typesetting, Ye Biao for photography of some projects, and Liang Ling's team for the translation of the Collection.

The Collection also uses project photos taken by other photographers to whom we are equally grateful.

图书在版编目（CIP）数据

持守本源·筑梦千里 广东省建筑设计研究院（GDAD）65周年作品集／陈雄，江刚 等主编．—北京：中国建筑工业出版社，2017.9
ISBN 978-7-112-21093-0

Ⅰ.①持… Ⅱ.①陈… ②江… Ⅲ.①建筑设计-作品集-中国-现代 Ⅳ.① TU206

中国版本图书馆CIP数据核字（2017）第192459号

责任编辑：唐 旭 李东禧 张 华
责任校对：焦 乐 王雪竹

广东省建筑设计研究院（GDAD）65周年作品集 编委会
主　　任　　曾宪川
主　　编　　陈 雄　江 刚　洪 卫　孙礼军
编　　委　　陈朝阳　李 巍　陈建飚　王业纲　罗赤宇　潘 勇　梁彦彬　黄 佳　周 文　李 鹏　陈 伟　李骏飞
　　　　　　陈伟雄　原效凯　郭奕辉　冯文成　蔡晓宝　郭伟杰　丁漫原　文 健　黄 毅　崔玉明　罗若铭　徐达明
　　　　　　黄伟勋　冯 伟　李振华　莫广英　许 滢　张 展　陈子莹　廖 雄　潘伟江　易 芹　彭祎环　陈宇青
　　　　　　谢 戈　邵 涛
执行编委　　丁漫原
封面题字　　陈宗弼
封面设计　　宋永普
图书设计　　廖荣辉
翻　　译　　梁 玲　鲍玉君

Editorial Board of The 65th Anniversary Collection of GDAD
Directors　　　　　　　　Zeng Xianchuan
Editors-in-Chief　　　　　Chen Xiong, Jiang Gang, Hong Wei, Sun Lijun
Board Members　　　　　Chen Zhaoyang, Li Wei, Chen Jianbiao, Wang Yegang, Luo Chiyu, Pan Yong, Liang Yanbin, Huang Jia, Zhou Wen, Li Pen, Chen Wei, Li Junfei, Chen Weixiong, Yuan Xiaokai, Guo Yihui, Feng Wencheng, Cai Xiaobao, Guo Weijie, Ding Manyuan, Wen Jian, Huang Yi, Cui Yuming, Luo Ruoming, Xu Daming, Huang Weixun, Feng Wei, Li Zhenhua, Mo Guangying, Xu Ying, Zhang Zhan, Chen Ziying, Liao Xiong, Pan Weijiang, Yi Qin, Peng Yihuan, Chen Yuqing, Xie G, Shao Tao
Managing Editor　　　　　Ding Manyuan
Chinese calligraphy on Cover Page　　Chen Zongbi
Graphic Design Cover Page　　Song Yongpu
Book Design　　　　　　　Liao Ronghui
Translators　　　　　　　　Liang Ling, Bao Yujun

持守本源·筑梦千里
广东省建筑设计研究院（GDAD）65周年作品集
陈雄 江刚 等主编

中国建筑工业出版社 出版、发行（北京海淀三里河路9号）
各地新华书店、建筑书店经销
恒美印务（广州）有限公司印刷
＊

开本：880×1230毫米　1/16　印张：21　字数：650千字
2017年9月第一版　2017年9月第一次印刷
定价：**198.00**元
ISBN 978-7-112-21093-0
　　　（30730）

版权所有　翻印必究
如有印装质量问题，可寄本社退换
（邮政编码　100037）